THE BOOK OF INDIAN DOGS

ALSO BY S. THEODORE BASKARAN

The Message Bearers: The Nationalist Politics and the Entertainment Media
in South India, 1880–1945
The Eye of the Serpent: An Introduction to Tamil Cinema
The Dance of the Sarus: Essays of a Wandering Naturalist
History through the Lens: Perspectives on South Indian Cinema
Sivaji Ganesan: Profile of an Icon
The Sprint of the Blackbuck (Ed)

THE BOOK OF
INDIAN DOGS

S. THEODORE BASKARAN

ALEPH

ALEPH

ALEPH BOOK COMPANY
An independent publishing firm
promoted by *Rupa Publications India*

First published in India in 2017
by Aleph Book Company
7/16 Ansari Road, Daryaganj
New Delhi 110 002

ISBN: 978-93-84067-57-1

1 3 5 7 9 10 8 6 4 2

This one is for Nithila and Sanjay

CONTENTS

AUTHOR'S NOTE

The two Indian breeds of dogs I raised at different points in time, one a Rajapalayam and the other a Tibetan spaniel (which I acquired from Tawang, when I was posted in Shillong, and who was with us for fourteen years), got me interested in the subject of indigenous breeds. Yet although I spent many years gathering material I never did manage to get down to writing a book though I did publish the occasional article. It was my friend, historian Thomas Trautmann of the University of Michigan, who finally persuaded me to start working on my cherished project. The pep talks he gave me when we visited him in Ann Arbor in July 2014 motivated me. His own book on the Asian elephant, *Elephants and Kings: An Environmental History*, had just been published and that was an inspiration. After I finished my first draft, Dianna Downing, a friend and a dog aficionado, went through the chapters and suggested improvements. It was a critical intervention.

The book is divided into an introduction and three sections. The introduction gives the reader a broad overview of when and how the domesticated dog arrived on the scene. Section I is a brief history of the dog in India, Section II describes the contemporary scene as it pertains to indigenous breeds and Section III is a guide to Indian dog breeds.

I am indebted to many friends and members of my family who helped me put the book together. Dr Malathi Raghavan of the University of Manitoba, a veterinarian, went through the appendix on stray dogs and offered useful suggestions. Conversations with friends interested in this subject, like Pune-based lawyer Meghna Uniyal, helped me to gain new insights. Job Thomas, art historian and lifelong friend, helped me collect material. Siva Siddhu of Tirunelveli, breeder of the Chippiparai breed, and with an impressive knowledge of the hunting hounds of south India, clarified many doubts. Whenever I called Dr N. Kandaswamy, former professor of Animal Genetics of the Veterinary University at Namakkal, with questions, he was always ready to talk. For a few years, when I was on the show committee of the Chennai branch of the Kennel Club of India, I learnt a lot from its secretary C.V. Sudarsan. I tested the validity of my insights by talking to Nithila, my daughter, and each time it was useful. She edited the photographs. Janaki Lenin provided some of the articles I needed. Rehan Ud Din Baber had uploaded on the Internet a number of documents relating to the subject of hunting and dogs and they were useful to my research. Venkata Ramana Pasila shared his knowledge of two rare breeds of Andhra and also gave me photographs. Sukhjinder Singh helped me with photographs. Anita L.G. Oechslin let me use a picture of a woman with a Caravan hound. Muthukrishnan and Arunkumar Pankaj let me use their photos of dogs featured in sculptures. I thank them.

Thilaka, my wife, a keen dog aficionado herself, was always there to egg me on whenever my enthusiasm flagged. I could not have completed the book without her encouragement.

S. Theodore Baskaran
Bangalore
January 2017

INTRODUCTION

Avoiding the friendship of those who resemble elephants,
seek the companionship of those who are like dogs;
for an elephant will kill his mahout whom it has known for
a long time,
but a dog will wag its tail even as the spear thrown at it is
still in its body.

Naladiyar (213) Tamil, circa 4 CE

In 1968, at a small wayside railway station in the extreme south
of India, near Tirunelveli in Tamil Nadu, I noticed a man waiting
for the train. With him were two exquisite white dogs. They lay
curled up on the floor of the platform, resting their muzzles on
their forepaws and looking listlessly at passers-by. Improvised aloe-
rope leashes were tied around their necks, the other ends held by
the owner who sat on his haunches, dragging at a bidi with great
determination.

They were Rajapalayam dogs, he told me, and they were on
their way to estates in the Anaimalais in the Western Ghats for
guard duty. He pointed to the many scars on the pale white coats

of the dogs and said that they had encountered quite a few boars in the hills near Rajapalayam. Sensing my scepticism, he said 'watch this' and clicked his tongue twice. The dogs were electrified into action. They were on their feet in a flash and stood pulsating with anticipation for the next command. That is how the dogs are put on alert when a boar is located, the man said. It was a magnificent display and I was fascinated. Would it be possible for him to procure a pup for me? By the time the train arrived, it was agreed that he would bring me one. I gave him my address and forgot about it.

Five months later, early one morning, he arrived at my Tiruchi home and handed over a small bundle. I removed the grimy towel covering it, and Madhu entered our lives. Almost immediately he began to look upon our house as his new home and soon assumed the role of guardian with all seriousness. My engagement with Indian dogs had begun in right earnest.

THE ORIGINS OF THE DOMESTICATED DOG

James Gorman, writing recently in the *New York Times*, observed: 'Before humans milked cows, herded goats or raised hogs, before they invented agriculture, or written language, before they had permanent homes, and most certainly before they had cats, they had dogs.' But the question that comes up is, since when? It is only around the 1950s that canine science took shape, and since then a number of studies have been done on the history of dog. Particularly in the last few years, there has been renewed scientific interest regarding the domestication of dogs or *Canis lupus familiaris*. Most of the studies on the subject have focused on the Grey wolf, the nearest and only ancestor of the dog. Scientists point out that 99.96 per cent of the DNA of dogs and wolves is common. The Wolf Science Center in Vienna, Austria, has been conducting long-term research using control groups of both wolves and dogs in large separate enclosures. The scientists there are seeking answers to questions such as: How did the dog develop into one of the most successful mammals in the history of nature? And, how did the dog arrive at its biological niche as a friend of humans?

In May 2015, the journal *Current Biology* published an article about a research project carried out in Sweden. Basing their analysis on an old fragment of the jaw of a dog, Swedish researchers came to the conclusion that the dog was domesticated 27,000 to 40,000 years ago. The bone which was studied was attributed to a period 35,000 years ago; it belonged to the Taimyr wolf, the most recent ancestor of wolves and modern dogs. Love Dalén of the Swedish Museum of Natural History said, 'Dogs may have been domesticated much earlier than is generally believed.'

But where did this domestication happen first? It was earlier held that the initial contact between humans and the ancestor of the dog, so dramatically described by Konrad Lorenz in his book *Man Meets Dog*, probably occurred somewhere in the Indian plains towards the end of the Mesolithic Age. As evidence, he points to the Indian wolf as the ancestor of the domestic dog. Certainly, some of the prehistoric paintings discovered in India depict scenes of hunting with dogs.

Ethologists have said that domestication probably began with a few wolves hanging around human settlements in order to scavenge scraps of food. Humans adopted them and later used them to hunt. There are scientists who differ with Konrad Lorenz about the period and place. In an article published in 2014 in *Science*, Olaf Thalmann of the University of Turku in Finland claims that this process of early domestication happened in Europe when humans there were still at the hunter–gatherer stage. He argues that the first 'proto-dogs' scavenged the carcasses left behind by these hunter-gatherers.

However, a recent research project at the University of California has come up with the view that the domestication of dogs occurred somewhere in the Middle East. This research team, led by B. M. von Holdt and R. K. Wayne, has analysed a large collection of wolf and dog genomes from around the world. They found that it was in the Middle East that the genomes of the two animals were most similar. They point out that some of the earliest dog remains have been found in the Middle East, dating back about 12,000 years. Publishing their research in the magazine *Nature* in March 2010, they suggest that wolves were first domesticated in the Middle East; later, when dogs spread to East Asia they crossbred further and more wolf genes were introduced into the dog genome. Other findings from this study show that dogs are able to take their cue from human body language which wolves are not capable of. This could have brought dogs closer to humans. They also note that the domestication of dogs and human settlement occurred almost at the same time—around 15,000 years ago.

In 1978, an archaeological excavation in Ain Mallaha in northern Israel uncovered physical evidence of one of the early bondings between humans and dogs. A grave that was 12,000 years old contained the remains of a woman and a pup. A related discovery occurred in Russia. In 2011, in Yakutsk in the Russian Arctic, the carcasses of two puppies, mummified in a landslide, and estimated to be 12,500 years old, were discovered. Scientists refer to these as the Tumat dogs, named after the village near which they were found. Dr Pavel Nikolsky of the Geological Institute in Moscow said that burnt bones were found near the site. He wondered whether this pointed out that dogs came to be attached to humans even before they became settled farmers. Research on the Tumat dogs is now going on in several laboratories around the world.

On 19 October 2015, a team of scientists presented their findings on the origin of the dog to the National Academy of Sciences in the US. Laura M. Shannon and Adam R. Boyko of Cornell University, along with a group of international scientists, studied the DNA of about 5,000 dogs from thirty-eight countries. (Their calculation is that there are about one billion dogs in the world, of which about 75 per cent are free-ranging and ownerless.) Their findings led them to Central Asia, including Mongolia and Nepal, as the place of origin of the domestic dog. This development is reminiscent of the manner in which genetic studies located the origin of modern human beings in East Africa. Though precise dating of the origin could not be done, scientists like B. M. von Holdt and R. K. Wayne surmised it was about 15,000 years ago. Between 3,500-3,000 BCE, dogs with collars appear. Archaeological sites of a thousand years later in Egypt show not only dogs with collars, but some with collars bearing their name—such as Brave One, Reliable, and so on. At Mohenjodaro of the Indus Valley Civilization (3,500-1,700 BCE) a terracotta figurine of a dog with a collar has been found.

A terracotta figurine of a dog with a collar found in Mohenjodaro.

The most ambitious research project on this subject is by Greger Larson, a biologist working in the archaeology department at the University of Oxford. He has been trying to coordinate the various projects that are on in different parts of the world on the origin of the dog. He is creating a vast database of the DNA of ancient dogs. It is reported that this project, which involves the Who's Who of dog researchers across the world, has a fund of $ 2.5 million from the Natural Environment Research Council in the UK. One of the objectives of the project is the detailed analysis of ancient canine DNA from bones obtained through archaeological excavations. R. K. Wayne, who is involved in the project, says, 'There is hardly a person working in canine genetics that is not working on this project.' Larson is hopeful that there will soon be an answer to the thorny question of the origin of the dog.

THE HISTORY OF DOGS IN INDIA

A statement attributed to Pliny the Elder (23-79 CE) says, 'Animals grow biggest in India. From India comes the dog that is larger than all others.' With its varied climatic zones, from the snow-covered mountain ranges of the Himalayas to the blistering deserts of western India, the Indian subcontinent is home to a bewildering variety of creatures. However, few people realize that this biodiversity is reflected in the country's canine population as well. India is home to a number of indigenous breeds of dogs. Sadly, some of them have already disappeared, due to indifference. This is unfortunate, especially given that in ancient times they were much prized around the world. Exported in large numbers (next only to the elephant where livestock exports were concerned) Indian dogs were used for hunting. Historians have recorded that Indian hounds were exported to Rome and to Egypt. Old travel accounts tell us that dogs from India were sent to Babylon. During the reign of Artaxerxes I, the king of Persia (465-425 BCE), four revenue-free villages were allotted to the Assyrian governor exclusively for the purpose of maintaining and breeding hunting dogs received from India. These were used for military purposes as well. When Alexander the Great invaded India and overpowered the local rulers, King Sopeithes of Gandhara (present day Gundurbar), gifted the invader 150 hunting dogs. Another story that has come down from this period says that to demonstrate the pluck of these dogs, two of them were set upon a lion. Even as one of the dogs suffered a badly injured leg, it held onto the lion. Alexander is said to have watched the display of tenacity with awe. The story may be apocryphal, but it indicates the mettle of the dogs. In his book titled *Indica*, the Greek writer, Ctesias (415-397 BCE), talks about an Indian tribe called Kynomologol that kept many large, ferocious dogs to protect them from wild animals.

Recent scientific evidence shows that long before these trade exchanges, Indian dogs might have travelled overseas. A study done in Adelaide, at the Australian Centre for Ancient DNA, has linked ancient Indian canine visitors to the dingoes of Australia. Researchers like Alan Cooper, director of this centre, say that the dingoes, which most closely resemble the Indian dog, probably arrived with some people from the Indian subcontinent around 4,300 years ago. He says that there is evidence of migration from mainland India to Australia at this time although the numbers may have been small.

◆

Although Indian dogs were in demand abroad, at home, except for some kings and nobles who indulged in hunting, the upper class and middle class shunned them. In fact, the dog was despised and the word 'dog' was used as a derogatory term in daily usage and in literature. A Tamil Saivite poet begs God in the *Thevaram* to 'have mercy on me, one who is lower than a dog'. The dog rarely figures in Indian mythology though one is aware that Lord Shiva,

Sculpture of Lord Shiva as Bhairava with his dog.

as Bhairava, had a dog as a companion. The dog was not associated with any ceremonial event (unlike the horse, the elephant, the camel or the cow) in any of the religions of India. The people who owned dogs and handled dogs were working-class people, farmers, graziers, trappers and hunters. It is clear that dogs in India were primarily work animals, like the horse and the ox, and the practice of keeping dogs as house pets seems to be of recent origin. They were not given the run of the house or treated as part of the family. The Tamil poet Kapilar records in a poem that the Brahmins did not allow dogs or chickens in their households.

Compared to other animals like the elephant and the cow there are relatively few references to dogs in ancient Indian literature and art, another indication of the lowly status dogs had in Indian society. In Tamil literature of circa 3-5 CE, we find some references to dogs, but mostly in the context of hunting. In the collection of poems called *Kurunthogai* by Sangam Age poet Sembula Peyaneerar there is a description of 'a big clawed hound, whose teeth shine bright like bamboo shoots'. Another poem in the same work talks about 'hounds that do not ever miss their prey'.

Some of the earliest representations of dogs in the Indian subcontinent are seen in prehistoric rock paintings, dating from about 30,000 years ago. In the Singanpur rock paintings in Madhya Pradesh, we see a barking dog rushing towards its quarry. With its straightened tail and exaggerated leg motion, the painter has tried to accentuate the speed and action of the animal. About 6 kilometres from Usilampatti in Tamil Nadu, in a place called Pudumalai, a team headed by archaeologist K. T. Gandhirajan found prehistoric rock paintings which included a hunting scene depicting a dog walking with a man. In some Stone Age sites in Tamil Nadu, we see depictions of hunting parties accompanied by dogs going about their work. This is clear evidence that dogs were domesticated as early as the Stone Age. These representations are, of course, stylized and do not give us any idea about the breed of the dogs.

A dog represented in a fresco of Ajanta.

Other early representations of dogs can be seen in the frescoes of Ajanta, which date from the second century BCE onwards. Kings of the Vakataka dynasty excavated from rock a series of caves in the valley along the Waghora River and decorated the interior walls and ceilings with frescoes relating to Buddhist subjects. There are at least three frescoes of the pre-Christian era in Ajanta that have representations of hunting dogs. In Cave I, the story of Janaka is told in the frescoes in comic-strip style. When his wife finds that he has eaten food discarded by a dog, she leaves him in disgust.

Another mural in Cave XVII featuring dogs tells a story from the *Mriga Jataka* about a queen who wants a golden deer. In a scene referred to as 'The Return of the King', the king sets out and captures a golden deer and returns with it in his chariot. Men with dogs on leashes, presumably from a hunting party, follow the chariot. In this panel, at least four similar-looking dogs can be seen, all on leashes. Another story is from the *Sutasoma Jataka*, a tale about Sudasa, the king of Varanasi, who sets out on a hunt with a pack of dogs. As a reflection of their respect for animals, Buddhist artists paid a lot of attention to detail while depicting these dogs in the Ajanta murals. They look distinctive—brown-coloured, short-eared, round-headed, and their tails are short. They are unlike the many Indian sighthound breeds that we see in show rings today.

Mughal miniature paintings from the sixteenth-century recorded aspects of the lives of the emperors. There aren't many representations of dogs in these works, as compared to elephants and horses. Even in their hunting scenes, we rarely see dogs. However, we read of a painter of miniatures named Manohar Das (whose work flourished between 1580-1620 CE and spanned the reigns of the emperors Akbar and Jahangir) who was given to depicting dogs. In a painting in the *Baburnama*, which depicts Khusrau Shah (the Indian emir who became general of the Mughal army and commanded the left wing of Emperor Babur's army in the Battle of Khanwa) paying homage to Babur, we see a Tazi dog, which is originally from Afghanistan. In another painting from Emperor Jahangir's time, two greyhounds feature. There is also a Mughal miniature painting which shows a beggar with a dog.

In a Rajasthani painting, dating back to 1707-1708, depicting Maharana Amar Singh II in his garden in Udaipur, we see two

Akbar with his hound.

hounds reclining. From their jewelled collars, it is evident that these dogs belong to royalty. Down south, the Nawab of Arcot, Saadatullah Khan I, sought dogs from Britain. In fact, in 1710, he presented six elephants to the officers of the East India Company and requested four dogs in return. The East India Company officials wrote to headquarters requesting 'three or four large Irish dogs and six pairs of swans to present to the Nawab', but I have been unable to confirm whether these were ever sent. We must note that most of the dog-fancying Indian royals did not show any interest in indigenous breeds.

I spoke to the novelist Amitav Ghosh about the opening of his novel *Flood of Fire*, which describes the British army on the march to India. He told me he had taken details from a lithograph of 1841 depicting such a scene. In these lithographs, called 'company paintings', we see dogs featured. In this particular etching, which shows the Bengal Regiment on the move during the Sind Operation under Sir Charles Napier, dogs are moving along with the soldiers. Some dogs are on leashes and some run free. A Banjara caravan with carts and camels also forms part of the movement, and so it can be presumed that these were the dogs of the Banjaras, which were used to guard their goods, brought along as supplies for the soldiers. The appearance of these dogs matches the description of this breed in *The Kennel Encyclopaedia* of 1908. Banjaras were also called Lambanis in some parts of India. Another lithograph of 1844, attributed to the artist Emily Eden, sister of Governor General George Eden, shows two soldiers of the Nawab of Awadh with two dogs which are clearly hounds.

After the advent of photography, by the middle of the nineteenth century, some pictures of dogs were made. What strikes the viewer is how different they are from present-day animals of the same breed. Take, for example, Rampur hounds in photographs taken at the turn of the century. In the photographs of Indian kings, zamindars, and British army officers, one is also able to see the dogs used for hunting. But for the captions provided for some of these photographs, we would not be able to identify them as Rampur hounds, since over the centuries cross-breeding has changed their appearance dramatically.

W. F. Sinclair of the ICS, writing in 1892, talks of nomadic herders called Thilaris in Maharashtra who had dogs of the same name. He describes the breed as follows: 'A tall, shaggy, lurcher-like dog, whose appearance suggests a cross between a greyhound and a black Newfoundland.' Some British officers, who used Indian hounds for hunting, have left informative notes about their experience with local breeds. R. W. Burton (1939), who raised a Rajapalayam hound at the end of the nineteenth century, recorded that dogs of this breed are 'good tacklers, sound feet, not too fast, stand the heat, splendid staying power and the best of constitution. They come from the Poligar country in the south of India'. He also owned Banjara hounds and said of them: 'Banjara dogs are very good but those of pure breed are seldom to be procured, as their owners, the most interesting nomad tribe of gypsy appearance and habits, will not readily part with them. These dogs hunt by both sight and scent and being fast and courageous are suitable for every description of hunting and coursing.'

◆

In the southern part of India, the early inscriptional references to dogs as companions to humans are found in the hero stones, often bearing inscriptions, that dot the countryside. It was the practice in ancient India to erect a stone plaque, now referred to as a hero stone (nadugal), to commemorate the valour of the person who had died in battle or fighting bandits. There are stones for dogs that died fighting a bandit or a wild animal, and even for horses and small pets like parrots. One of the earliest dog memorials is in a tiny village called Eduthanur, in the Tiruvannamalai district of Tamil Nadu. Here there is a memorial stone in the temple of Oomaivedipan erected in memory of a young man named Karundhevakathi and his dog called Kovivan. They fought thieves who were trying to

A hero stone in Eduthanur, Tamil Nadu.

rustle cattle and were killed in the process. The relief sculpture in the stone depicts the man in profile, wielding a dagger in one hand and a bow in the other. The dog, positioned behind him, is shown baring its teeth. It is a stocky animal, with a well-rounded head and short bat-like ears. The inscription on the stone narrates the story and reveals that it was erected in the 34th regnal year of the Pallava king Mahendravarman I (580-630 CE). This, incidentally, is one of the early references to a dog with a name. There are other such hero stones for dogs, such as a sixteenth-century one in a village called Thachanpudur in the same area.

Similarly, in Anantapur district of Andhra, in the village of Phalaram Gollarahatai, a king of the Nolamba dynasty (ninth century CE) erected a memorial stone for his hunting hound that died tackling a boar. In the village of Lingala in Kadapa district, a memorial stone stands to celebrate the memory of a dog called Porakukka who gave his life when his master, a soldier named Vikramatiyan, was attacked and killed.

Another commemorative relief sculpture of a dog is exhibited at the Government Museum in Bangalore. This large memorial stone, 2 metres tall and 1.5 metres wide, which bears an important epigraph, was discovered by the British archaeologist B. Lewis Rice in 1889. Realizing the significance of the inscription, he brought the monument to the museum where it occupies a prime place at the very entrance. This stone memorial, which dates back to 950 CE, celebrates the memory of a hound that was the favourite of Manalera, a commander in the army of Rashtrakuta king Krishna III (939-967 CE). The nineteen-line inscription in Kannada on the stone tells the story, in a poetic manner, of the hound Kaali, which the commandant was given by the king himself in appreciation of the valour he demonstrated during a decisive battle against the Cholas at Thakkolam, a village in Tamil Nadu. Later, during a hunt in the forest, in a place called Belathur, Manalera was attacked by a boar and wounded. Kaali, who was with him, took on the boar and killed it. But in the process the dog was mortally wounded. Heart-broken, Manalera buried the dog in the Chellalingeswara Temple in the village of Aathagur in Mandya district and erected a hero stone opposite the temple. He even donated a paddy field for the upkeep of the memorial. A priest named Gorava was appointed to offer worship daily at the memorial. Manalera made the priest promise that he would not eat until after he completed the worship every day. On the upper portion of the hero stone is the sculpture, done in deep relief, featuring the dog hanging onto the snout of the boar. Kaali was a stocky dog, with small ears, short muzzle and short tail. Below the sculpture is the inscription telling the story of the death of the dog and of the grant of land.

A very early sculptural representation of a dog in India is found on the Gangadhara panels of the Pallava period (seventh and eight centuries CE). There is a story in the Ramayana about King Bhagiratha who undertook a punishing penance to pray to Ganga to come down to earth, which was reeling under severe drought. In answer to his supplication, Ganga came down in such a torrent that she would have destroyed the earth. It was at this point that Lord Shiva intervened and broke the mighty force of the river in his matted locks of hair. This story is referred to as the Gangadhara episode, and it appears in a number of representations in Pallava art in which Lord Shiva is represented as Gangadhara.

What is of interest to us here is that invariably there is a figure of a dog somewhere in the Gangadhara panels. It is very prominent in the large panel in the seventh century CE Pallava rock-hewn temple in the Tiruchirapalli Rock Fort. A seated dog is featured on the upper portion of the panel. Michael Lockwood and other scholars point out that the Puranas are silent about any dog in this episode. The mystery of the dog in the Gangadhara panels

A sculpture of a soldier with his dog from the Chola dynasty.

continues.

In the Chola temple at Gangaikondacholapuram, built in the eleventh century CE, there is a miniature sculpture depicting a fierce-looking dog held on a taut leash by a soldier. The dog sports a broad collar. On one of the panels of the frescoes inside the Brihadeeswarar Temple in Thanjavur is a painting showing a near replica of this type of dog. Since it is part of a mural, the colours are shown, revealing it to be white-coated. In the same temple, in a sculpture narration of the story of Kannappa, a hunter and devotee of Lord Shiva, he is shown accompanied by four dogs.

In the monuments belonging to the Vijayanagar period (1336-1565 CE), there are a number of sculptural representations of dogs. Most of them are present in hunting scenes. In Hampi, in Karnataka, the capital of the Vijayanagar empire, there are sculptures depicting a full range of royal activities, including the hunting of deer with the help of dogs. Hunters with leashed dogs, done in relief sculpture, adorn the sides of the Mahanavami platform, overlooking the royal centre. King Krishnadevaraya (1542-1565 CE) would ascend the platform during the Mahanavami festival, where he would perform worship and view the parades of animals, musicians and dancers, mock battles and fireworks below.

In Srivaikuntam, in Tamil Nadu, in the Vaikundanathar Temple (belonging to the Nayak period which followed the Vijayanagar period) there is a sculpture depicting a lively scene of a hunter and his wife. She is removing a thorn from his sole, while he stands on one leg, with a boar slung across his shoulders. Beside him there is a dog.

In Madhya Pradesh, 16 kilometres from the town of Indore, stands an unusual temple, one erected for a faithful dog. Local legend has it that the temple centres around an itinerant trader, a gypsy named Rewa Banjara. In dire need of money at one point, he mortgaged his dog to a local banker. The gypsy got the money, left his dog with the banker and went on his way. One day, bandits broke into the bank and looted money and jewels and buried their loot in the riverbank. The dog took scent, led the banker to the spot where the bandits had hidden the money, and began digging there. On recovering his lost money, the banker decided to release the dog. He wrote a note to the gypsy trader and tied it to the dog's neck. The dog ran to join his master. Upon seeing the dog, the gypsy thought that it had escaped from the banker. In a moment of anger, he drew his sword and killed the dog. It was only then that he noticed the letter attached to its neck. He was overcome by remorse, and as atonement built a temple over the grave of the dog. This tenth-century edifice, known as Kukarramath Rinmuktheshwar (kukar means dog in Sanskrit) attracts devotees who believe that a visit to the shrine will free them from debts. The Archaeological Survey of India, which has declared this a protected monument, attributes the temple to the Kalachuri period of King Kokalyadev I, as the architectural style corresponds to this period (twelfth century CE).

Before I end this brief historical section on the dog in India, I should mention the graves for dogs that belonged to famous figures in history. Chhatrapati Shivaji's dog, Waghya, probably the best-known dog in India, was as famous as his master's horse, Kalyani, and has a memorial in Raigad Fort. Legend says that when the king died in 1680, his loyal dog gave up his life by jumping into the funeral pyre. The story may be apocryphal, but the dog's fame spread, and in 1936 a statue of the dog sitting on his haunches

Memorial of Waghya, Shivaji's dog.

was placed on top of a 5-metre-high platform erected on the spot where the king was cremated.

In the historic town of Bidar in Karnataka are two sixteenth-century tombs, one in Shiva Nagar and the other in Guru Nagar known as Kutte ka Kabur. These are pointed out as the tombs of two dogs of the king of Bahmani. The tomb in Guru Nagar is a more elaborate structure. In Mahuli, there is a memorial for King Shahu Maharaj (1874–1922) with a statue of his dog Khandya on top. In Ootacamund, guided tours take visitors to the grave of District Collector John Sullivan and his dog.

Visitors to the cattle farm at Hosur in Tamil Nadu are shown the dog cemetery. This farm was started by Tipu Sultan to raise horses for his cavalry and to breed Hallikar bulls that were used to pull his gun carriages. When the British took over, they named it

Cemetery for dogs in Hosur, Tamil Nadu.

the Army Remount Depot. It was in the hands of the military till 1938 when it came under the animal husbandry department of the state government. In those days British army officers were in charge of this farm. The dog of the superintendent of the farm died in 1924, and is buried under the massive banyan tree in front of the official residence, with a tombstone bearing an inscription. Since then subsequent superintendents also buried their dogs here, in a row, with similar tombstones. The last dog to be buried was Rani, a Pomeranian that died in 1972.

These are some instances of dogs being honoured in the past, too infrequently in my opinion. Our dogs are an important part of our heritage and our lives and need to be given the respect and importance that is their due.

SECTION II

THE CONTEMPORARY SCENE

Certain common characteristics can be observed in all Indian dogs, especially those from the plains. Most of them are outdoor dogs and are oriented towards speed and action. Therefore, they become stressed if tied up or kept in a confined space. From historical times onwards, they were bred to course on the plains, chasing quarry like blackbuck, chinkara and hares. As the emphasis was not in making them showpieces, the practice of docking (cutting the tail) was unknown. Even dewclaws were not removed at birth. Similarly, castration was rarely resorted to except in the case of sheepdogs, which were allotted to a flock. In recent times in south India, owners who do not want the bloodline of their dogs to be diluted have resorted to castration. Indian breeds were not trained as gun dogs or as retrievers, although, as we have seen, a number of them were used to run down quarry.

Most of them are lop-eared to protect the delicate inner tissues of the ear when they operate in thorny and bushy terrain. Almost all the Indian breeds are sighthounds, that is to say, they hunt by sight. So they have keen eyesight.

Understandably, the coats of the dogs of the plains, like the Rampur hound, are short and smooth, to suit the hot climate. So there was very little need to groom them. In fact, grooming as a practice was unknown among the owners of these dogs. Dogs were never bathed since they were not meant to be shown and were meant for work. Even in the last few years, when some of these dogs have been brought to the show ring, they are not groomed. These dogs need minimum maintenance and they have fewer medical problems compared to foreign breeds. Veterinarians testify that genetic problems such as epilepsy are usually absent among these breeds and instances of single-testicle (monarchism) or undescended testicles are rare. Traditionally, before the advent of modern veterinary medicine, the breeders used Siddha medicine to treat the dogs.

There is nothing special about the food they are fed. I have spoken about this to a number of dog owners and they all say that they feed millets to their dogs. Beef, although available in rural areas in some Indian states, is not fed to the dogs. As mutton is expensive, most owners settle for cooked rice or ragi. A few of them also feed their animals goat's milk. The dogs get to eat some meat when they are out on a hunt. When animals like hares and monitor lizards are degutted, the offal is given to the dogs.

Often when hunting dogs are in forest areas, they face a threat from leopards which have a partiality for dogs and are known to go for the throat while attacking. So a device known as a 'leopard collar' is put on the dog. It is made of light metal like tin or thick leather.

Another distinctive Indian practice when it comes to indigenous breeds has to do with the branding of dogs. I have seen dogs bearing distinct branding marks on their hind legs. Some of the branding patterns are very similar, and it is clear that the branding was not done to prove ownership as is sometimes claimed. The instrument usually used to brand is a steel spoke from a bicycle wheel. It is heated till it is glowing red and pulled over the skin of the dog. The owners explained this as a method of treatment for certain kinds of illnesses. Some of them believe that this procedure gives general protection to the animals. A veterinarian told me that the possible effect of this might be to activate the immune system. Some dog owners in rural areas believe that the brand marks neutralize the effects of the 'evil eye' that may be cast on the animal. Most of the branded dogs I have seen were Mudhol hounds and a few Caravan hounds.

◆

The breeds of dogs that evolved in different geographical areas—well-adapted as they were in diverse environments—remained undiluted until three centuries ago. This is one of the reasons for the sturdiness of these dogs. The gene pool was intact until the colonial period, when dogs from other lands came to be imported into India and interbreeding began. The first to import dogs were the Portuguese in the sixteenth century. But it was only after the British tightened their hold on India that there was a steady inflow of dogs. There is a record from 1850 in which A. H. A. Harvey says: 'Hundreds and thousands of these dogs (hounds) were sent to India at enormous expense, to Madras particularly.' More and more hounds were brought in from the Middle East and from Europe, and these were bred with the native breeds.

When we look at the representations of dogs in the frescoes of Ajanta or in the other historical sculptures and paintings, we notice that the features of the dogs portrayed are very different from what we see in the country today. The square head gives way to a long, pointed muzzle. The stocky frame changes into a deep-chested, well-sprung body. It is evident that the breeds have undergone a change in recent centuries. We may get a clear picture about the breeding dynamics when the DNA studies on indigenous breeds, which have just begun, are completed.

We have reliable records about imports of dogs from Europe during the colonial period. This was related to the growing popularity of hunting among British officers in India. The abundance of wildlife, sustained by varied habitats, offered them tempting opportunities to hunt. The hordes of migratory waterfowl that sought out and landed in water bodies in the country attracted duck shooters and they needed dogs trained in retrieving. Indian breeds, which were able to chase and corner quarry, were not useful as gun dogs nor were they good at retrieving. European breeds like retrievers and pointers, which had been bred and trained as gun dogs, were imported into India. We know that bloodhounds were brought into India during the time of the Portuguese in the sixteenth century. By the seventeenth century, archives show that other European breeds had arrived on the subcontinent. We have it on record that Emperor Jahangir requested Thomas Roe to get him different breeds of dogs from England. Roe did bring British mastiffs to the emperor who provided caretakers along with silver vessels for these dogs. As I have mentioned earlier in the book, in 1710, the Nawab of Arcot requested the East India Company to get him three or four Irish dogs (hounds). There are other mentions of imported dogs in accounts of the time. A bulldog owned by an English chief of a storage place on the west coast killed a sacred bull, and the angry crowd that gathered there massacred nearly eighteen Britons. The monument to the slain officers, John Best and seventeen others, bears the inscription: 'They were sacrificed to the fury of a mad priesthood and an infuriated mob.' In *The Rifle and the Hound in Ceylon*, written in 1854, planter S. W. Baker records having brought in a dozen greyhounds from England, for the sole purpose of hunting. Major E. Napier's book written in 1840, *Scenes and Sports in Foreign Lands*, records that horse traders from Arabia brought Arab greyhounds with them into India where they were sold for fifty to two-hundred rupees.

Surprisingly, the import of greyhounds persists till today. Recently, the *New York Times* carried a story about greyhound racing at Faridabad in Punjab in which hounds imported at enormous cost from Canada, England and Ireland participated. It is a sign of prestige for the landlords of Punjab to own greyhounds and participate in the event. In October 2016, Mudhol hound racing was held in the town of Mudhol with ₹11,111 as the first prize. Unconnected to this practice, in a recent move, the Government of India has banned the import of foreign breeds for breeding purposes. This has been done through an order of the Director General of Foreign Trade.

◆

As the Raj extended its shadow over India, the number of Indian rajas importing foreign breeds increased. A good example of this was the Nawab of Junagadh, Mohbet Khan Rasul Khan, who owned 150 dogs including bull terriers and Border collies with a separate kennel and caretaker for each dog. There was a veterinarian, an Englishman, to take care of them. When the nawab

Nawab of Junagadh with his dog.

wanted to see any one of the dogs, it was brought to him in a palanquin. The nawab was fond of hunting and he owned a pack of hounds, which he used for that purpose. Oil portraits of his favourite dogs, that he commissioned British painters to paint, are still exhibited in Junagadh Palace. They wore diamond-studded collars. One painting shows a marriage between two of his dogs, the celebrations of which lasted for three days (including one day which was declared a state holiday). The nawab invited Lord Irwin to one such event, but the Governor General politely declined the invitation. It is recorded that a number of the nawab's vassals attended the wedding in full ceremonial dress. After the partition of India in 1947, as he was fleeing to Pakistan on a Dakota plane which was filled with his wives, treasure and dogs, one of his four wives on board found that her child had been left behind. She asked the nawab to wait and rushed to the palace to get the child. He filled the vacant seat with two dogs and fled.

◆

Once the wave of imported dog breeds arrived in India, cross-breeding was inevitable. Since many of the imported dogs could not stand the heat of the plains, some dog fanciers began to cross them with local breeds in order to have dogs that were tough and could withstand the heat. The gene pool that had remained almost inviolate for centuries got diluted.

Some British dog fanciers, however, did try and preserve the uniqueness of Indian breeds, particularly that of Himalayan dogs. Understandably so, because of all Indian dog breeds, dogs in the Himalayas had grown closest to humans, and had been given the run of the house, a privilege the dogs of the plains did not enjoy. In fact, the sheepdogs, which were mostly in the Himalayas, were much closer to humans than the hunting dogs of the plains, which were kept separately and taken out only on hunting trips. The concept of ritual pollution through dogs was unknown among the people of the hilly region. There was another reason for the British attraction to Himalayan dogs. With their bushy coats, these dogs resemble certain European breeds. Their thick coats enabled them to withstand the cold. This meant they could tolerate the English weather. Consequently, some of these breeds were exported to England. They were given English names like Tibetan terrier and Tibetan spaniel. Soon an association of owners of Indian dogs was founded in England with 2,000 members.

◆

In India, the British interest in dogs manifested itself in the form of dog-fancier clubs. In 1896, the North Indian Kennel Association, a precursor to the Kennel Club of India, was formed in Lahore, where it held its first dog show. In 1900, it was affiliated with the Kennel Club of London, and in 1908 it became the Kennel Club of India with Lord Minto, the viceroy of India, as its president.

Reverend Herbert Wheeler Bush, who spent nearly two decades in India, from 1881 onwards, in Delhi and Jullundur (present-day Jalandhar), was one of the first Britons to take a keen interest in Indian dog breeds. Reverend Bush, while talking about the Banjara breed in *The Kennel Encyclopaedia,* said: 'Eastern sportsmen have never yet, within the memory of Englishmen, set themselves to consistently manufacture a breed for special purpose.'

Oxford-educated Reverend Bush served in India in various places such as Delhi, Dalhousie and Sialkot. For a few years he was the principal of Lawrence School in Murree. In due course, he earned the reputation of being an expert on 'Eastern and Oriental hounds'. The notes he left about some of them, including one on the Afghan hound he raised, gives us an idea of the scene at the time. He was a key figure in founding the Kennel Club of India and in drawing attention to Indian breeds. He founded the *Kennel Gazette*, which is still published every month with impressive regularity from Chennai as the *Indian Kennel Gazette*. He died in Britain in 1927.

Gradually, clubs were founded in metropolitan areas and dog shows were held. Beginning in 1952, shows were held in Ootacamund. However, no sustained interest was shown in local breeds of dogs until the 1960s. Until then, even prominent Indian leaders were usually only interested in imported breeds. In photographs Dr B. R. Ambedkar is seen with a pair of fox terriers. Down south, E. V. Ramasamy Naicker's favourite breed was the dachshund. Jawaharlal Nehru had a honey-coloured golden retriever whom he had named Madhu.

◆

By one account India has about four million pet dogs which is much less than the eighty-three million in the United States. However, in India, the number of dogs being kept as pets is growing at the rate of 15 per cent a year. Pet clinics and the pet goods trade are also increasing. Bangalore has seen about fifty dog spas (where owners bring their pet dogs for bathing and grooming) coming up in the last three years. According to another estimate, there are 1,432,522 pet dogs in Bangalore. But very few of them are indigenous dogs.

The concept of breeds, as it is known today, was not common in the subcontinent until recently. People used a variety of expressions to distinguish between the various kinds of dogs. In Tamil, for example, it was common to use the word 'caste' (jathi) in place of 'breed'. I think that this is true of other languages as well—in the 1990s, when I was taking my dachshund for a walk one morning in Ahmedabad, a curious youngster approached me and asked in English, 'Uncle! What caste is your dog?' This might be amusing but even in countries where breeds have been recognized for much longer it was only in the last two centuries or so that the idea of breeds and breed characteristics took firm root, as a consequence of the formation of associations and clubs devoted to dogs.

In India, some dog enthusiasts argue that Indian breeds are landraces and not breeds. What is meant by this is that these animals belong to a specific area and are raised by a specific community or tribe. Within the 'breed' sizes and colour vary. In recent decades, we have moved away from such rough and ready methods of classification and an attempt has been made to standardize Indian breeds, especially since the Kennel Club of India began to allow Indian breeds to be shown in the ring and insisted on certain standards being met.

However, as there has been no real attempt to establish a methodical and scientific way to classify and set standards for the various breeds, there is little standardization of characteristics within breeds. This has led to all sorts of confusion. M. Krishnan, the great naturalist, recounted one such story from the 1940s about a Rajapalayam dog he got from a breeder in southern Tamil Nadu. When the pup, which he named Chokki, grew up she stood only 56 centimetres in height and weighed a meagre 18 kilograms. Krishnan wrote to the breeder asking whether the sire was a true Rajapalayam and got the following reply. The breeder pointed out that the size of the offspring did not invariably depend on the parents and added, in a postscript, that he knew men five-feet-nothing in height but it had never occurred to him to doubt their humanity.

It is a sad commentary on the regard we have for our own

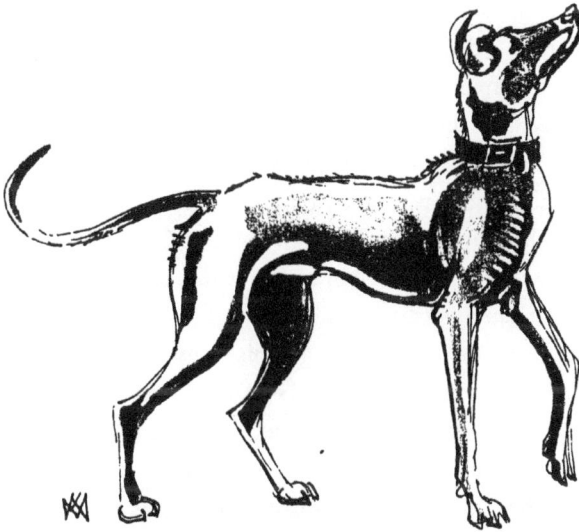

A sketch of Chokki by M. Krishnan.

native dogs that their breeding has been neglected for centuries now. We have seen how the bloodlines of Indian breeds were diluted by European breeds. Also, in more contemporary times, the craze to acquire foreign breeds for pets has exacerbated the neglect of local breeds, especially as far as the middle class is concerned. To make matters worse, they are even being neglected by traditional hunters and people in the rural areas because of the ban on hunting by the Wildlife Protection Act 1972 and the emphasis on wildlife conservation. For many owners in the rural areas of Tamil Nadu, for instance, the main reason for keeping dogs was hunting, and this was indulged in mostly as a hobby. Hunting is part of the rural culture of Tamil Nadu along with rooster fights, jallikattu (bull-vaulting) and rekla (cart race). In 2011, in Sivaganga, Tamil Nadu there was an attempt to introduce dog racing with indigenous breeds but it did not catch on. The last race was held in 2014 in which thirty-four canines participated. The home of many of the fabled breeds of the region like Kombai and Chippiparai is along the foothills of the Western Ghats in the western part of Tamil Nadu where there were communities that eked out a living by selling bush meat acquired through hunting. One of the memories of my childhood in a village in Tamil Nadu is the arrival of trappers at our doorstep hawking hares and partridges. The other traditional work done by these dogs, like guarding and keeping watch, have become largely redundant due to modern surveillance measures and equipment like CCTV cameras. The only Indian breeds that have thrived are Himalayan breeds such as the mastiff, the Lhasa apso and the Tibetan terrier because of interest in these dogs in Europe and America where, as we have seen, there are organizations for their preservation.

Some Indian breeds have competed at prestigious international dog shows such as the Crufts Dog Show, the biggest in the world in which 25,000 dogs compete. A few Indian breeds have won the Best in Show title. In 1984, a Lhasa apso won this coveted title. In 2007, it was a Tibetan terrier that won and recently in 2012 a Lhasa apso was once again chosen the Best in Show. However, what is to be borne in mind is that not a single Indian breed has been recognized as a distinct breed by the Fédération Cynologique Internationale, the apex body which governs classification. Based in Brussels, it recognizes 343 breeds from eighty member countries including dogs from Indonesia (the Kintamani Bali dog), the Taiwan dog and the Japanese Akita.

Neglect of Indigenous Breeds

While writing this book, I was saddened to find that many legendary Indian breeds are on the point of vanishing or have already disappeared. In the eighteenth century, a Frenchman travelled around the country and identified as many as fifty distinct Indian dog breeds. These included breeds like the Lut, a blue or fawn coloured dog, which is no longer found. The breed Alangu, from the Thanjavur region of Tamil Nadu, and the Malaipatti dog from the Sabarimala area in Kerala, have vanished. Other breeds such as the Bhallar dog from the Himalayan foothills, the Gazelle hound, the Mahratta hound and the Shencottah have not been seen in living memory. Even our best-known breeds like the Rampur hound or the Kombai have not received the attention that is their due. Likewise, a breed from Manipur called Tangkhul hui, a stocky medium-sized dog, with bob tail and bat ears, remains uncared for.

There are many reasons for the gradual disappearance of local breeds. The primary reason, as we have seen, was the arrival of exotic breeds of dogs during the colonial period due to which the local breeds suffered neglect. Moreover, there was little interest on the part of the government to preserve the various breeds that existed in the country. What was true of the government of British India was also true of Indian rajas. As we have noted, few

Indian royals were interested in indigenous breeds. Out of the 565 kingdoms in British India, only a few like the Nawab of Rampur (Rampur hounds) and the Raja of Kolhapur (Mudhol hounds) cared for them. Even this limited patronage for local breeds was reduced when the princely states were abolished following the independence of India. Rajas had so many other pressures to deal with during that transition that dogs were neglected.

More recently, as I have noted, other developments such as the ban on hunting led to the neglect of Indian dog breeds, especially in the rural areas. The dilution of bloodlines due to interbreeding with foreign breeds has hastened the decline of Indian breeds.

Before I end this section, I should mention an interesting discovery about canine blood transfusion. This came about, curiously enough, during the Indian Peace Keeping Force (IPKF) operations in Sri Lanka (1987-1990). The Indian Army had taken along a squad of forty-two dogs, consisting of Labradors and Alsatians. Spending long stretches of time in bunkers and trenches, the dogs contracted 'tick fever' (ehrlichiosis) which led to anaemia and they badly needed transfusions of blood. The Veterinary College in Chennai had established a canine blood bank and the infected dogs were brought in two at a time to Chennai. They were administered blood transfusions and were saved. Dr G. Baranidharan, blood bank officer, who has written a guide on canine blood transfusion, says, 'Dog blood groups are identified based on the surface antigen of the erythrocytes (DEA) and there are more than eight blood groups identified till date.' After testing many breeds he concluded that Chippiparai is a safe and universal donor. In the rest of the world, the greyhound is considered a universal donor. The blood bank in Chennai, which boasts state-of-the-art equipment, is carrying on advanced research in this area and is testing more Indian breeds to check if some of them could be safe donors. On World Blood Donor Day on 14 June 2016, two Chippiparai dogs in Chennai, Mani and Rusty, who have been donating blood regularly to this bank, were honoured at a function in the college.

Attempts to Revive Indigenous Breeds

Unlike other Asian countries like Japan (the Akita, the Shiba Inu and the like) and China (the Shar Pei, Pekinese, pug and shih-tzu), where indigenous breeds have been bred with care and dedication, as I have shown, we have been woefully neglectful of our canine heritage. There have been only sporadic attempts, both individual and institutional, to revive indigenous breeds. This neglect has persisted for over a hundred years. By the 1960s, a few individuals became interested in Indian breeds and worked for their recognition. The first opening came when the Kennel Club of India allowed some of them, such as the Rampur hound, to be brought into the show ring. With these arose the issue of breed standards. In the early 1980s, some additional breeds were shown in the rings. In 1981, in the Kolhapur Canine Dog Show, forty Caravan hounds were shown. By 1984, standards were established and at least ten breeds—including the Rampur hound, Caravan hound, Rajapalayam and Himalayan mastiff—were accorded recognition in the country.

Gomathy Srinivasan, a minister in the Tamil Nadu government of 1981, was an Indian breed fancier, and when the portfolio of animal husbandry was allotted to her, she set up the Dog Breeding Unit for Indigenous Breeds in Saidapet in Chennai. Functioning under the management of an assistant director, it began with

Rajapalayam, Kombai and Rampur hounds. The centre functioned well for a few years, but with a change of government, it fell into decline. When I visited the unit in 1985 it was in poor condition. It was evident that being caged and without any exercise the health of the dogs had deteriorated. The enclosures in which the dogs were kept were cramped and filthy. Despite this, there was a waiting list of 130 for pups. In 2012, the Tamil Nadu government woke up to the unit's existence and ₹30 lakh was allotted for its improvement. Unfortunately, more and more exotic breeds such as Labrador retrievers were being bred there, as a result of which Indian breeds were neglected. That year about fifty pups were sold.

Sometime later, People for the Ethical Treatment of Animals (PETA), with permission from the Animal Welfare Board of India, inspected the unit and found the conditions appalling. It gave an adverse report to the Animal Welfare Board, which directed the state government to close down the unit but the government did not close the unit. So PETA took the case to the Madras High Court in 2014, seeking its closure. The Tamil Nadu Kombai Dog Revival Foundation also approached the high court, pleading that the unit be allowed to continue and the court passed the following order. 'The native dog breeding unit should correct all lapses pointed out within a period of three months and restore its original object for which it was established and the Animal Welfare Board will inspect it after three months and file a report. If the defects and acts constituting cruelty continue the shelter will be shut down.' However, the unit continued to function though the standard has not improved in any way. On 4 December 2016, the Madras High Court ordered the breeding unit to be closed.

Karnataka's attempts to nurture its indigenous breed, the

Mudhol hound, is a good example of what can be achieved if there is the will to preserve indigenous breeds. The Mysore Kennel Club took the initiative to bring indigenous breeds into the show ring and work towards their preservation. They began with Mudhol hounds, the well-known local breed of Karnataka, and allowed them to be shown in dog shows and participate in championship competition. Since 1995 Mudhols have been seen in show rings in the state and this has increased the popularity of the breed outside its native tract of Bijapur district. At a dog show conducted by the zila parishad in Mudhol about 350 hounds were shown. Since 2014, the Mysore Kennel Club has been conducting the Indian Breeds Speciality Show in Bangalore.

The most impressive and sustained effort in this direction is being taken by Karnataka at the initiative of Govind Karjol, who was minister of social welfare in the state government in 2011. The Karnataka Veterinary, Animal Husbandry and Fisheries Sciences University (KVAFSU) showed interest in the Mudhol breed, which began to be considered part of Karnataka's heritage. The state government allotted five crores of rupees and an area of 40 acres for the purpose of breeding the dog. The centre began functioning in 2011. In 2015, there were twenty-eight hounds in the centre, and it began sending dogs to shows in various towns in south India. In addition to the work in the centre, the staff go out to the field, providing free vaccination and health check-ups. A separate state-of-the-art whelping centre has been built on campus.

Microchipping of dogs is also carried out for free. In India, microchipping started in 1998. Microchipping is a process in which a chip is inserted under the skin of the dog, between the shoulder blades. It contains a number and details of the breed and owner. This forms part of a database. The primary purpose of microchipping is to identify the breed and its owner.

With the help of the four researchers in the centre, a survey was carried out in the region to identify Mudhol breeders. Seven hundred of them were identified and registered in a database. This became the Mudhol Dog Breeders Association. A family that has two dogs and two bitches is considered a breeding unit, and a cash incentive is given to the breeder. Veterinary doctors regularly visit these families and advise them on how to breed and take care of the animals. Training programmes for breeders are also conducted.

The Indian Council of Agriculture Research has come forward to support this idea, as it was seen as a means of livelihood for people in the rural areas, although all its other schemes are for cattle. The Mudhol hound is looked upon by the council as a companion animal, and the breeders are given medical and financial help. Scheduled caste and tribal people are given incentives to breed these dogs, and earn money by selling pups which are certified as thoroughbred Mudhols. The Karnataka Veterinary, Animal and Fisheries Sciences University at Bidar has taken an interest in the dog breed and in 2009 set up the Canine Research and Information Centre (CRIC) in Thimmapur village near Mudhol town. The centre certifies dogs, installs microchips on the animals and maintains records.

Besides these attempts to care for the Mudhol hound, a group of dog lovers in Bangalore who were concerned about indigenous breeds got together in 2013. Led by Dr B. C. Ramakrishna they formed the Society for Indian Breeds of Dogs. It has been registered with the state government and is affiliated with the Kennel Club of India. The membership fee is ₹2,000. The society intends to promote breeds by holding special shows and through conducting DNA studies. An important project the society has in mind is to work for the care of native breeds in Afghanistan, Nepal and Myanmar, as the breeds in these countries have a common ancestry with Indian breeds. It is to the credit of this society, that along with the Kennel Club of India it conducted the first ever Indian Breed Speciality Dog Show in Bagalkot in Karnataka in 2014. The event provided a fillip to the efforts to save indigenous breeds and more shows came to be organized. These shows attract dog fanciers and breeders and kindle interest in them. Judges from other countries observe Indian dogs in the ring. Andrew Brown, a hound expert from Australia, was a judge at the Indian Breeds Dog Show in Nagercoil, Tamil Nadu, in 2015. He said to me that he felt the breeds in southern Tamil Nadu 'have all the qualities of a sight hound'. In this show a Rajapalayam and a Pashmi won in the group category of hounds. The other hounds shown were the Caravan, Mudhol, Kombai and Chippiparai.

In Secundarabad, a few dog fanciers, led by R. Upender Reddy, decided to salvage some of the remaining Indian breeds in that region and formed the Ethnic Indica Canine Society in 2002. The society began its activities in September 2002 by conducting a survey of indigenous breeds in the Hyderabad area of Andhra Pradesh. Reddy gathered a team of dog enthusiasts and undertook an extensive survey of four breeds that were found in that area: the Caravan, Pashmi, Pandikona, and Jonangi. With the help of Raghunath Patil, a dog enthusiast, who breeds Caravan and Pashmi dogs, a rural dog show was organized on 30 December 2002 in the village of Janwal, located about 30 kilometres from Latur, and considered the birthplace of the Pashmi breed. Nawab Nazeer Yar

Jung and G. V. N. Krishna Rao, experienced judges of the Kennel Club of India shows, served as judges for the show. Exhibitors from Nanded, Bijapur and Osmanabad brought around 110 Caravan hounds to show. The opportunity was also utilized to educate the owners about breeding and caring.

In March 2005, the Chennai chapter of the Kennel Club of India launched a project to revive the Rajapalayam breed. Symbolically, a pair of pups was presented to the secretary of the club at the dog show that year. The plan was to breed along scientific lines and propagate the breed by selling the pups. The kennel, located on a farm on East Coast Road and under the leadership of C.V. Sudarsan, has continued its efforts in nurturing indigenous breeds. In May 2015, it conducted a camp at Nagercoil for registering indigenous breeds of dogs. Word spread, and a number of dogs were registered. To encourage participation, the registration fee was waived. The dogs were microchipped and the owners' details were documented. This camp was held as a prelude to a special Indian breeds show to be held at the end of 2016 at Kanyakumari. A similar camp was held in Rajapalayam in April 2016 where documentation and microchipping of indigenous breeds of dogs was done.

There are private breeders in Tamil Nadu who have taken up promoting Chippiparai, Kombai and Rajapalayam dogs on a commercial basis. For example, there is a dog farm, named David Farm, in a small town called Marandhai, near Tirunelveli. This farm, which had sixteen dogs in 2015, is managed by John Arthur, a dog enthusiast, who sells a pup for around ₹8,000. There are similar breeders, selling Mudhol and Caravan pups in other places.

A few non-governmental organizations, with the help of government researchers, assist in the revival of some breeds. SEVA in Madurai, working in the area of indigenous knowledge and livestock, along with the National Biodiversity Authority based in Chennai, has been showing interest in Indian breeds, and it chose Pon Elangovan, a Rajapalayam dog breeder, for the Breed Saviour Award in 2013. Sarvodaya Sevabhai Samstha, another NGO, has been working in Sarnath in Uttar Pradesh, with funding from Help Animals India.

In a pioneering attempt, the National Bureau of Animal Genetic Resources (NBAGR), based at Karnal in Haryana, began making efforts to set standards for Rajapalayam and Chippiparai breeds. The team led by Dr K. N. Raja visited areas of Tirunelveli and Madurai in 2014 and recorded details from breeders. In the field they noted the external features and body measurements of these two breeds, including information on the colour of the coat, weight, and also information as to which community was rearing them. They followed the twenty-five-point standard set by the American Kennel Club to define a breed. The idea was to study them using microsatellite (DNA related) markers. Unfortunately, the Indian Council of Agricultural Research, under which NBAGR functions, discontinued this project.

In recent times, besides the various initiatives I have mentioned, it is heartening to see that there is growing interest among Indian dog fanciers in indigenous breeds. B. Kolappan of *The Hindu* writes perceptive articles on indigenous breeds. Social media has had an important role to play in this regard. Groups working for the preservation of certain breeds are active on Facebook. Dog owners are able to exchange notes and breeders are able to keep in touch with prospective buyers by announcing litter arrivals. Photographs of pups and dogs are uploaded. In fact, to prove the authenticity of the breed of the pups, some breeders even publish photos of their dogs copulating. Facebook has become a popular platform for debate about breed characteristics and nomenclature. Magazines, during show season, occasionally carry articles on indigenous breeds although it is unfortunate that they repeat the same matter every year. Television channels, like NDTV Goodtimes, also feature programmes on Indian dogs. However, much remains to be done. Attitudes need to change. To give just one example of the obstacles that stand in the way of the large-scale revival of interest in our canine heritage let me end this section with a story about how Indian dogs are commonly viewed. In 1974, a few Indian dogs, including a Chippiparai, were inducted into a police squad in Tiruchirapalli. One of the dogs helped in nabbing a thief from his hideout by following the scent trail. When the dog was produced in court, the judge would not accept its 'findings' as evidence because it was a mere 'country dog'.

A GUIDE TO INDIAN DOG BREEDS

As we have seen, the material on dogs from the historical period in India is scanty, but there is adequate evidence to show that they were basically used as work animals—for hunting, guarding and in herding sheep. Of these, hunting seems to have been the most common work into which dogs were pressed. Many of the nomadic and pastoralist communities in India had dogs with them, often unique breeds. For instance, the Kalbelia tribal people who roamed the Great Thar Desert had a breed of dog called the Kalbelia (after the tribe). Another breed that some pastoralists of Rajasthan owned was known as the Thilari—this has now disappeared.

Depending upon local needs, dogs could be trained to perform unusual tasks. The Jonangi breed, for example, was used to herd ducks in the marshlands of Andhra. This is similar to the use of Newfoundland dogs in places like Pennsylvania in the United States to assist fishermen with gathering up nets and the catch in lakes. As has been noted, Indian breeds were not trained to be gun dogs or retrievers. Even when they were used by British officers, Indian hounds were used only for chasing and holding the quarry at bay.

Naturalist M. Krishnan, who was working in the princely state of Sandur in what is now Karnataka, provides a graphic description of how Indian hunting dogs were used. He was accompanying a group of men and dogs who were out hunting boars.

> We kept skirting the foot of a long hill, and at daybreak the dogs suddenly began to tug at their leashes and led us up the hillside diagonally; they never faltered or changed direction but went straight across the shoulder of the hill in a beeline to a hollow in which the pigs were lying up. When their excited whimpering told us the quarry was near, the dogs were slipped and streaked in a pack into the hollow. The men positioned themselves lower down where the pigs would break cover and very soon there was a rushing sound as of a minor avalanche and some forty pigs burst out of the hollow. The men scattered to allow free passage to the main body and then closed in on a small juvenile pig and speared it, and the hunt was over.
>
> Afterwards I carefully estimated the distance from that hollow to the point on the periphery of the hill where the dogs had picked up the pig scent and it was at least a kilometre. True the wind was in favour of the dogs coming straight across the hillside and keeping low to the ground, but still I was impressed by their scenting ability.

In Tamil Nadu the dog that leads the pack on these hunting trips is called kathu nai, meaning it is able to catch the scent well and follow the trail. Hunting expeditions are risky and some of the dogs die in the encounters. M. Krishnan has written a moving article about tending a dying dog that was wounded by a boar.

The Thotinayakka community in Tamil Nadu specialized in

hunting with dogs. Each household had three or four animals. In Erumaipatti, a small tribal hamlet at the foothills of Kolli in Tamil Nadu, I had an opportunity, in 1992, to observe a traditional hunt with dogs. At dusk one of the hunters stood at the edge of the village and blew the kombu, a brass horn. I noticed that none of the dogs were tied up. Dogs that lay curled up in various corners of the village as if in a stupor sprang into action and gathered around him. The other men in the party leashed the dogs and they all moved towards the foot of the hills. Some of the men carried spears made of bamboo with metal tips. When a dog scents a boar, it strains on the leash and is let loose by the handler. The men follow the dog and as it locates the hideout of the boar, they deftly spear one when the sounder breaks cover. If the party bags a boar, they return to the village singing, thus announcing the success of the hunt. Most of the time they are successful and bag at least a few hares, if not a boar.

Hunting dogs were very much part of the life of farmers and others. One of the living traditions in Tamil Nadu is to offer terracotta votive offerings of cattle figurines to village deities like Ayyanar, seeking good health for domestic animals. Similarly, small terracotta figurines of dogs were also placed in the shrine of Ayyanar by those seeking protection for their dogs. I have seen them in a number of temples of village deities.

In Tamil Nadu, two types of hunting forays were in vogue until recently: one was called thangal vettai (camp hunt) in which a single owner with a few dogs camped in the forest for a few days and conducted hunts. In the other, called koottu vettai (combined hunt), a few dog owners collectively took their dogs on a hunt. During these outings, the huntsmen degutted the hares caught and offered the entrails to the dogs.

Different methods are used to keep the dogs fit. In some places the pups are put in a pit and as they repeatedly jump to get out, they develop a well-sprung chest. Chippiparai dogs are often taken for a swim if there is a water body close by.

Dogs were often used in ceremonial hunts organized as a part of festivals to honour local deities in rural areas in Andhra Pradesh and Karnataka. The Ethnic Indica Canine Society at Hyderabad has recorded the proceedings of one such hunt near Bijapur that took place as part of Ugadi (Telugu New Year's Day) celebrations at a Durga temple in 2003. Groups of men with their dogs, mostly Pashmis and Caravan hounds, set out at the crack of dawn to hunt hares. When a dog caught a hare, it would not let it go, and the hunters would have to yank it from its mouth. With the passing of the Wildlife Protection Act in 1972, forest officials clamped down on this practice. But often exemptions are sought for the hunt as a religious ceremony. Wildlife NGOs have tried to help the state stop this custom by pointing out to the organizers the damage the practice does to wildlife. In some places it has worked. But in remote localities, the tradition goes on, albeit stealthily.

◆

From the Himalayan slopes, along which the Gujjars move with their flocks of sheep to the plains of Tamil Nadu, herders and farmers have raised sheepdogs. However, the principal home of the Indian sheepdog is the Himalayas. They not only had to herd the flock but also protect them against predators like the big cats. Recalling his memories of the 1920s, Jim Corbett, who spent much of his life in the Garhwal area of Uttarakhand, wrote in his book *The Man-eating Leopard of Rudraprayag*:

> These big, black and powerful dogs that are used by packmen throughout our hills are not accredited sheepdogs in the same sense that sheepdogs in Great Britain and Europe are. On the march the dogs keep close to heel, and their duties only start when camp is made. At night they guard the camp against wild animals—I have known two of them kill a leopard—and during the day and while the packmen are away grazing the flock they guard the camp against all intruders. A case is on record of one of these dogs having killed a man who was attempting to remove a pack from the camp it had been left to guard.

The pastoralists of western India, like the Rabaris, who move along with their livestock looking for pastures, have sheepdogs with them. Though the association between herders and dogs goes back to antiquity, the sheepdogs of India are usually not trained for mustering or grouping. They merely guard the flock. However, Gujjars used sheepdogs to drive flocks from place to place and keep sheep from straying, in addition to guard duties.

One of the earliest accounts of Indian sheepdogs is by Lieutenant General W. Osborn who wrote about them in the *Journal of the Bombay Natural History Society* (1892). He pointed out that in India sheepdogs were not trained as meticulously as in England, but were merely raised with sheep. While hunting blackbuck near Bellary, Osborn came across sheepdogs that chased and killed a buck he was pursuing. From his description of the dogs we can infer that they were Pashmis.

In another instance, he writes about the sheepdogs he came across in Bidar (now in Karnataka) in 1863. An inspector, a Scotsman, working on the railway line being laid across the Morna River, shot and killed two sheepdogs when a pack of them attacked him. The case came before the Deputy Commissioner and the Scot had to pay a large sum of money as compensation to the owner of the dogs. Osborn became curious about the high value placed on these dogs. He located the owner and asked him about their training. The shepherd explained the process to him. When they wanted to train a new dog, they would carefully choose a male pup from the litter of one of the larger breeds. The pup would then be attached to an ewe that had lost its lamb. The pup would be suckled by it. It took about three weeks, said the shepherd, to get the pup used to the wet nurse. The pup was not weaned early so that it would be stronger. By maturity the dog would be totally at ease with sheep and vice versa. When it attained adulthood, the dog would be castrated. The idea was to prevent the dog from associating with other dogs and restrict its loyalty only to sheep.

In Tamil Nadu, pastoralists would make an enclosure at night to shelter their sheep. This pen was called patti and the dog would usually stay the night there. These dogs were referred to as patti nai or the dog of the patti. These dogs were also castrated when they reached maturity. Interestingly, the word for dog in Malayalam is patti.

◆

Dog fanciers disagree on the precise number of indigenous breeds to be found in this country. As I have mentioned earlier in the book, a French traveller identified as many as fifty distinct breeds in the eighteenth century. Major W. V. Soman, in *The Indian Dog*, the first major book to be published on indigenous breeds of India, about half a century ago, listed about sixteen breeds. In my considered opinion, based on over four decades of personal observation, research and reading, there are not more than twenty-five distinct indigenous Indian dog breeds to be found today. These are described in the pages that follow. The guide has been organized into three groupings—working dogs, companion dogs and hounds. Within each grouping the dogs are listed alphabetically.

WORKING DOGS

The dogs that I have collected together under this grouping are used for a variety of purposes by their owners—herding, guarding property and even sometimes to hunt. However, none of these breeds are hounds or gun dogs (in the sense of dogs that are used to point or retrieve). All the Indian breeds that are hounds are grouped separately.

Bakharwal

The Bakharwal is a mountain dog originating in the Pir Panjal mountain area of the Hindu Kush and the Himalayas. It belongs to the Asian molossers variety. It is a sturdy animal, standing nearly 70 centimetres tall and weighing about 40 kilograms. A typical mountain dog, its furry coat and plumy tail give it a majestic look. It looks like a smaller version of the Tibetan mastiff. Its coloration is usually tan or black. Grazier tribes like Gujjars and Bakharwals use this dog to guard sheep and hunt. The dogs are also referred to as Gujjar dogs.

In 2009, when the Tribal Research and Cultural Foundation based in Poonch, Kashmir, conducted a survey on the Bakharwal, it was observed that only a few hundred of these dogs remained. An appeal was made to the Ministry of Environment and Forests and Climate Change to protect them. These dogs are greatly fancied in cities, and so male dogs are taken away at a good price. This affects breeding in the native areas. Added to this is the fact that a Bakharwal bitch is slow to breed. She litters once a year, often with only two to four pups.

Himalayan Mastiff

The Himalayan mastiff, one of the largest Indian breeds, is the iconic guardian dog of the Himalayas. It is also known as the Tibetan mastiff but can be found in many parts of the Himalayas, including Sikkim. Standing 70 centimetres tall and weighing 75 kilograms, it is a majestic looking animal along the lines of other well-known mountain dogs such as the St Bernard or the Great Pyrenees to which breeds it is related. The massive head, curled-up tail and luxuriant mane are characteristic of the breed. The coat is shaggy, like the sheep it often guards. The colour is generally black, but tan and honey coloured animals also occur. The tawny patches over each eye are referred to as 'eyes of darkness' by the villagers, who believe that these dogs can spot the devil. I have seen it only once in its home terrain in India, in a village in Sikkim, with a grazier and his flock of sheep.

This dog has been reputed to stand up to leopards and wolves in defence of sheep. Legend has it that these dogs were used for military purposes. They were trained to terrify horses and their ancestors were reputed to have fought the hordes of Attila the Hun and Genghis Khan.

There is another dog which looks like a smaller version of the

Himalayan mastiff, and is sometimes called the Bangara mastiff. In fact, the local people do not use this name, which was given to this breed by Major W. V. Soman in his book on Indian dogs. It usually has a black-tan coat and is distinctly smaller than the Himalayan mastiff. There is a variation of this breed called the Bearded Himalayan mastiff or Kinnauri kutta, which is a favourite with shepherds. This dog is found in the Kinnaur region of Himachal Pradesh and is often described as a separate breed. However, it is in fact a variation of the Himalayan mastiff.

The Himalayan mastiff is an ancient breed. When Marco Polo travelled to Tibet in the fourteenth century, he wrote with poetic exaggeration that the dogs he found in Tibet were as big as donkeys. The invasion of Tibet by the British in 1903 brought news of Tibet's dogs to the outside world. Through an article in *The Kennel Encyclopaedia* (1908), Reverend H. W. Bush spread information about these Himalayan breeds. In the beginning of the twentieth century, a number of mountaineering teams from Britain began their expeditions to Mount Everest and they did not fail to notice the varieties of dogs in the mountain country and left notes about them in their reports.

Occasionally these dogs were used as beasts of burden to transport bags of salt across the mountains. When the Eastern hillmen came down as far as Kolkata to sell their merchandise, they left these mastiffs behind to guard their women and homes. The bark of this mastiff is deep, like the gong of a bell, and it carries a long distance in the stillness of the mountains. They are sluggish by day, but come into their element by night, so a prospective buyer chooses his dog after nightfall. In a strange ritual meant to instil courage, the tip of the pup's tail is cut, roasted and fed to it. Of course, one cannot say with confidence that this actually makes the dog brave, but the fact is the Himalayan mastiff has a great reputation for pluck and is credited with the ability to stand its ground. Traditionally, the patronage this breed received from successive Dalai Lamas has enhanced its reputation. These dogs are a favourite in dog shows in India.

In the 1900s this dog was taken to Britain, and in the 1906 Crystal Palace Dog Show a single specimen was shown. Later, the breed was established in the United States where there are at least 300 animals now. Recently, the newspapers reported that a rare golden-coated Tibetan mastiff was sold by a breeder for $2 million in the Zhejiang Province of China. Owning a mastiff is a status symbol in China.

In 1985, at the Delhi Dog Show I met Don Messerschmidt, an anthropologist from the United States, who was doing field research in Kathmandu and had come to show Kalu, his Himalayan mastiff. Along with his research in anthropology he was also studying Himalayan dogs and eventually published his findings in a book called *Big Dogs of Tibet and the Himalayas*. He describes a few hitherto unknown Himalayan breeds in this book, including the Sha-kyi or the hunting dog.

Himalayan Sheepdog

Another big dog of these mountains is the Himalayan sheepdog that is used for guarding cattle and herding. Occurring in Ladakh and Nepal, it is also known as the Bhotia or Bhote kukur. This breed received the special attention of King Mohidant of Meerut, and in recent years the Raja of Dumraon has taken a keen interest in these dogs and has been breeding them. In 2014, the Himalayan sheepdog made news when one was used to run errands between outposts for Indian soldiers in the snow-bound Siachen area of the Himalayas. Though it is a rare breed, it does make an occasional appearance in dog shows. It is a large dog with a plumy tail, standing around 60 centimetres tall and weighing between 30 to 40 kilograms. This dog sheds its coat once a year. Sleeping most of the day, it guards the flock in the night. This breed was one of the four that was featured in a set of Indian postage stamps released in 2002.

Jonangi

This comparatively small-sized breed, standing only 30 to 40 centimetres tall and weighing around 20 kilograms, is found in the Godavari Basin of Andhra Pradesh, particularly around the Kolleru Lake area. The wrinkles on the face of this loose-skinned dog are distinctive. The coat is very short, with a sheen that recalls velvet cloth. It comes in varied colours including brindle and black. Its place of origin was a vast marshy area around the lake. At one time these dogs were able to catch crabs and fish and live on them. Since these dogs are used to traversing marshy stretches, duck breeders use them to herd ducks. Achyutha Ramayya of Tanuku village is one of the breeders of this dog and works at promoting this breed. The Ethnic Indica Canine Society team, which included Nawab Nazeer Yar Jung, visited his farm in October 2002 and saw eight pairs of dogs in his kennel. The Nawab collected a few pups and is trying to propagate the breed. One of its distinct characteristics is the very short coat, through which the skin is visible. The ears are held erect, and the tail is short and curved. Two of these dogs were shown in the ring in a dog show in Mysore in 2014. However, the Kennel Club of India has not yet accepted the Jonangi to be shown. Jonangi is one of the pristine indigenous breeds.

Kombai

This is a legendary breed of south India, originally occurring around the Cumbam-Uthamapalayam tract in Madurai district. There is a village in this area called Kombai which was a seat of a poligar. However, M. Krishnan is of the opinion that it is not the village Kombai that lends its name to the breed. The word 'kombai' denotes a fertile, low-level tract in the scrub to be found in the foothills of the region and as this breed originated in this sort of terrain Krishnan was of the opinion that it had been named after it. Tough and tenacious by temperament, the Kombai is sometimes compared to a bull terrier. There is much variation in size. The height can be between 38 to 80 centimetres and the weight from 20 to 30 kilograms. The muzzle and the tip of the tail are black and the body is usually tan in colour. Rather stocky in build, it sometimes has a dorsal line of fur running down the back and growing in the opposite direction to the rest of the hair on its coat (reminiscent of the Rhodesian ridgeback)—Kombai fanciers look for this feature as a mark of a thoroughbred. There are occasional specimens with bat-like ears but usually the ears droop partially, and only the tip of the tail is curved. Kombai aficionados say these dogs mature late.

The Madura Country, written in 1868, by the British civil servant James Henry Nelson, describes some of the qualities of this breed. Apparently, local rulers, the poligars, fancied them so much that they would, with alacrity, exchange a horse for one. One peculiarity of this breed is that when provoked it will not be subdued by superior force, as the larger Rajapalayam can be. The Kombai is primarily a guard dog and is not meant for hunting, and although it is sometimes called a 'bearhound' the reference is probably to its fabled fighting qualities—it could even take on a bear. Colonel James Welsh, the British East India Company military officer who fought the poligars in southern Tamil Nadu, and who documented the operation in his book, *Military Reminiscences* (1830), says that the fort of the Marudhu brothers in Kalaiyarkovil near Tirunelveli was defended by a brace of fierce Kombai dogs. In 1972, the Southern Railways dog squad in Tiruchirappalli had a Kombai that was used to patrol the goods yard.

In the Ramanathapuram area is a variety of dog called Mandai and it is much fancied. It is a stocky animal, on the primitive lines of the Kombai and comes mostly with grey coat though other colours are also seen. Many breeders think the Mandai is only a variation of Kombai; it is also referred to as Ramanathapuram Kombai.

The status of this breed remains obscure, as it has been diluted due to indiscriminate breeding to meet heavy demand. However, some dedicated dog enthusiasts have joined hands to set up the Kombai Dog Preservation Unit and are making an attempt to resuscitate the breed.

Koochee

The Powinder people, a nomadic tribe in Afghanistan, used Koochee dogs for herding. With the disappearance of the nomadic lifestyle, the prevalence of this breed has diminished and it is now almost gone. W. V. Soman wrote that merchants from the Northwest Frontier Province, who used to come as far as Bombay to sell their goods, introduced the dog to India. In the olden days when they used to travel with camels and cattle, Koochees were used to guard these caravans at night. It is a large, stocky dog, standing about 70 centimetres tall and weighing around 50 kilograms.

A few enthusiasts have joined hands to revive this breed. The Bully kutta, another large dog which is found in Pakistan and Afghanistan, and was used in dog fighting in the olden days, is a close relative of the Koochee. This breed has fallen into disrepute as it is being used in the blood sport of dog fighting. Though banned, such fights are still held clandestinely in some places in Punjab and Haryana.

Sindhi

Sindhi dogs were brought into western India from the Sind region by horse traders. Its original home is the desert area of Rajasthan and the Sindh region of Pakistan. It looks like a cross between a mastiff and a greyhound. These heavy-boned dogs were also used as sheepdogs. Until a few decades ago these dogs could be found in Gujarat and Rajasthan, particularly in the Nara Valley of Mirpur Khas district. The King of Danta, a principality in Gujarat, is said to have patronized the breed and used it for pig-sticking, a sport which he enjoyed. The Sindhi is a large, smooth-coated dog, standing 70 centimetres tall and weighing about 40 kilograms. The colour of the coat is usually solid brown, red or brindle.

Pandikona

Aficionados of the Pandikona breed trace its origin to the Vijayanagar Empire (fourteenth to fifteenth centuries CE) of which the village of Pandikona was a part. After the Battle of Talikota in

1561 CE, in which the Vijayanagar army was routed, the breed and the area in which it was to be found (it is now in the Kurnool district of Andhra Pradesh) were neglected. The Pandikona's home turf was once the hunting ground for the aristocrats of the region and the dog was originally used to hunt boar (pandi in Telugu means pig). In 2002, the Ethnic Indica Canine Society team visited this village and was able to find six healthy Pandikona dogs that they took for breeding.

Stocky in build, it is a medium-sized dog, measuring between 50 to 66 centimetres. The Pandikona shows wide variations in size as the local breeders never chain their dogs and are not given to selective breeding. It is smooth-coated, with colours varying from solid fawn, shades of cream, white to black with white patches, brindle being the rarest. It has bat-like ears, which are often cropped by the owners. The black rim around its eyes is considered distinctive. Some of these dogs bear marks of branding. This breed has not been tampered with and is pristine. Pandikona are used for guarding and hunting. They are very faithful and good with children.

Patti

The Patti dog, the popular farm assistant of the peasants of the Kongu region in Tamil Nadu, was originally used by villagers to guard the patti, the pen into which the sheep were herded for the night for protection against wolves and foxes. The Patti dog's alertness is proverbial. The term patti nai, however, is commonly used to refer to all the dogs that guard the patti. This seems to be the only variety in this region that can be called a sheepdog. The dog has bat-like ears. The colour is usually tan but can vary widely. It is a hardy breed and can subsist on meagre rations. It stands between 45 to 55 centimetres tall and weighs about 25 to 28 kilograms. This breed is so varied that it is commonly regarded as a landrace rather than a breed. Certain nomadic tribes still breed them.

Pati Patia

The Pati patia breed found in Mayurbhanj and Korapet districts of Odisha is used by the tribals of the region for tracking prey and hunting. It is a medium–sized dog with a sharp snout and half-folded ears. The dog's coat is brown or black, and some animals develop a distinct saddle. Slightly smaller than a Labrador, the Pati patia is around 50 centimetres in length and weighs about 25 kilograms. In 1983, some of these dogs were trained for police work in Odisha such as guarding. But the status of the breed seems uncertain; for this reason it is the only 'breed' included in the book that is not illustrated. I'd be interested to hear more about this dog.

COMPANION DOGS

Companion dogs have been around for centuries. Usually very small in size, companion dogs have been bred to be lapdogs, and are usually to be found in homes, although occasional breeds may be used as guard dogs within the house. Some of the best-known companion dogs, such as the Pekinese and the Shih-tzu, originated in the imperial courts of China hundreds of years ago. For a long time afterwards companion dogs were to be found exclusively in the homes of the wealthy. However, today companion dogs are very popular pets, especially in urban areas where they are comfortable in small flats. All the Indian companion dogs are of Tibetan origin but are widely distributed across the Indian Himalayas. They were taken to other parts of the country by Tibetan refugees who were resettled in states like Karnataka. The best-known Indian companion dog is the Lhasa apso. Less popular are the Tibetan spaniel and the Lhasa terrier.

Lhasa Apso
Tibet is an excellent breeding ground for dogs as they are considered sacred there. Tibetans believe that souls of erring priests enter

dogs, and the lamas encourage dog breeding. The fabled Shih-tzu originated here. So did the Lhasa apso, a low-slung, hairy, tiny dog standing just 28 centimetres tall and weighing no more than 7 kilograms. Because of its thick coat, it needs a lot more care than other breeds to keep it tidy and free of parasites. The dog prefers to be clean and is miserable when its coat is untidy. So periodic grooming is necessary. It is the best known of the Himalayan breeds and was one of the earliest to catch the attention of Western dog fanciers. Through the officers of the Indian Army, and later through Tibetan refugees who were settled in Karnataka, the Lhasa apso has become popular as a companion dog all over India.

It is an old breed that has evolved naturally, unlike many European dogs that have been cross-bred. In Tibet, it was considered an honour to be presented with this dog, which was believed to bring good luck to its owner. It earned the name 'Bark Lion Sentinel Dog' (Abso Seng Kye) from the Tibetan aristocracy, which used it as a watchdog in their homes. In the Himalayas, while the Himalayan mastiff guarded the exterior of monasteries, it was the tiny Lhasa apso that was kept to guard the interiors. This breed in fact prefers to be indoors. It is often trained to rotate the Buddhist prayer wheel in monasteries and hence acquired the name 'prayer dog'. It was a tradition for the Dalai Lama to present one of these dogs to visiting dignitaries. Though this breed originated in Tibet, it is found all over the Eastern Himalayas, including Sikkim, and Arunachal Pradesh. In 1985, I saw these dogs in the Monpa villages near Bomdilla. When the British colonized India, they took a particular fancy to this native breed. It was taken to England in the 1930s where it became popular as the 'Talisman Dog' and was soon established there. From Britain this breed found its way to the

United States where there is an association of Lhasa apso owners. I saw some excellent specimens at a dog show in the Piedmont region of North Carolina.

Tibetan Spaniel

Smallest of the Himalayan breeds, this dog stands just 25 centimetres high, weighing about 6 to 7 kilograms. It has a dense, silky coat and is coloured red, black or fawn. It sheds hair once a year and needs regular grooming. Its original home is the district town of Jomsom and the Chumbi Valley in Tibet from where it spread to other Himalayan regions. It is a misnomer to call it a spaniel but the early name given by the British has stuck. Locally known as Simkhyi, it resembles the Japanese spaniel in appearance, with its short muzzle and pendent ears. The plumed tail, which is the hallmark of this breed, is held high over the back and is especially attractive. Its feet are hairy, to help in walking on snow. Cynologists refer to this feature as 'rabbit-foot'. Dogs originating in the higher reaches of the Himalayas have this feature. Locally referred to as Jemtse apso, these dogs, like Lhasa apsos with whom they share a common ancestry, have been trained by Buddhist monks to rotate prayer wheels in monasteries. So they too are sometimes called 'prayer dogs'. The Tibetan spaniel also acts as a hot-water bottle for its human companions who tuck this cuddly little dog under their garments as they move about inside the monastery. Its courage is well-known, and it is sometimes referred to as a miniature mastiff, having gained a reputation as a fine sentinel.

The British were very fond of this breed. Claude White, the first political officer in Sikkim, had a number of these dogs in his home. According to *The International Encyclopaedia of Dogs*

(1974) the first Tibetan spaniel was taken to England by Mrs Frank Wormald in 1905, and in the 1920s a medical missionary, Dr Nancy Grieg, took more of these dogs with her when she left India. But it was only after World War II that the breed grew in popularity in England. Sir Edward and Lady Wakefield took a bitch named Lama to England and in 1947 procured another by the name of Dolma. With these two they began breeding the dogs, and by the early 1970s the number of Tibetan spaniels in Britain had risen to a thousand. However, this breed is not very popular in India, and I am not able to figure out why. One hardly sees them in dog shows here.

I raised a female Tibetan spaniel, Tashi; I picked her up from Tawang in Arunachal Pradesh when I was working in Shillong. She was with us for fifteen years. Aloof and distant, she was independent and had a distinct character. She was very hardy. Even when she was hit by a car and suffered a ruptured liver she pulled through where a less tough animal may not have made it.

As with the Lhasa apso, this breed was made available to Indian owners by Tibetan refugees who settled down in this country.

Tibetan Terrier

This heavy-coated, medium-sized dog stands about 35 to 40 centimetres tall and weighs around 12 kilograms and comes in various colours. The long hair on its face has earned it the nickname 'the bearded dog'. This is an ancient breed that is locally known as Tsang apso, meaning the dog from the Tsang region in Tibet. This dog can be seen around Dharamsala in Himachal Pradesh where it is often used as a watchdog and a mascot. It often makes an appearance in the show ring, particularly in Delhi and Kolkata dog shows. It is one Indian breed that is popular in dog shows around the world.

Along with the Tibetan spaniel, this dog became popular in England. The coat of the Tibetan terrier is clipped and blended with yak hair to produce a soft, semi-waterproof cloth—a unique use of dog hair. The average life span of this dog is around twelve years. The dog sheds its coat and requires careful, periodical grooming.

HOUNDS

All indigenous hounds are sighthounds which track and hunt their prey with their keen eyesight as distinct from scent hounds, which use their noses to follow scent trails. Some of the Indian hounds, such as the Rajapalayam, are reputed to have an excellent sense of smell, but they too are primarily sighthounds.

Alaknoori

This breed is a close relative of the Caravan hound. H. H. Shahu Maharaj, the king of Kolhapur, was keen on pig-sticking. For this purpose he imported greyhounds from Europe and kept them in his private hunting ground in the village of Alakanoor in Karnataka. Though these hounds were known for their speed, they reacted adversely to the heat of Peninsular India. The Maharaja bred these hounds with Caravans. Through repeated attempts he succeeded in creating a breed that was christened Alaknoori, after the village in which it originated. It was described as a dog with a short, white coat with black patches and pink ears. But the breed was not taken care of and seems to have petered out due to indiscriminate cross-breeding. I have not seen even a single specimen of this breed and would be interested to know if it still survives in this country.

In June 2016, a book written by a prolific writer on dogs, Judy Taylor, titled *Alaknoori Training Guide* on handling this breed was published in the United States. It gives details about taking care of this dog, from the house-training stage onwards.

Banjara Hound

'The indigenous canine aristocracy is not large. But it exists. Among the best breeds are the hounds kept by the Banjaras, a caste of half gypsy carriers and traders', wrote J. H. Kipling (father of Rudyard Kipling) when he lived in Bombay in 1894. The nomadic tribe of Banjaras moved with their merchandise on pack animals, and they always had these hounds accompanying them to guard their settlements wherever they camped.

One of the early references to this rough and rugged hound is in the *Journal of the Bombay Natural History Society*. W. F. Sinclair, of the Indian Civil Service, based in Thane, made a note in 1892 referring to these nomads as Wanjari or Lambani and to their dogs as Wanjari hounds. He compared the dog to a Danish boarhound and said it was of fierce temperament and not amenable to discipline. Here is his account of the dog:

> They were very fierce and brave, and were kept in order chiefly by force: though not all dangerous to their friends. One old Wanjari lady once reduced a dog that attacked me to order by throwing her skirt over his head and sitting down on him. These dogs had fine short coats, commonly black or mostly black, but sometimes fawn or brindle. I never saw or heard of one of this race in the possession of any one

but a Wanjari and even among that caste they were not very common.

Reverend H. W. Bush, who as we have seen was one of the original authorities on Asian breeds, referring to the Banjara hound in *The Kennel Encyclopaedia* (1908) said, 'Undoubtedly they must be recognized as a distinct breed for they breed true to type in every way.' He points out that even in the 1860s this breed was noticed by dog enthusiasts. They had noticed that the Banjara hound resembled the Persian greyhound but was not so slim. The British officers who used these dogs to hunt recorded the remarkable courage of these hounds.

Some breeders feel that the Banjara hound is just another name for the Caravan hound. The dog stands around 65 centimetres tall and weighs 25 kilograms. It has a silky black coat mottled with grey or blue.

Caravan Hound

Found in the southern districts of Maharashtra and in the Deccan Plateau, this breed is also known as Karwani. Some dedicated breeders in Udgir, a small town in Maharashtra, have saved the breed from going extinct. There are also breeders in the village of Mhalangi in Latur district of Maharashtra. According to these breeders, Caravan hounds are descendants of Afghan and Saluki breeds, which arrived in the country with Arab traders and their camel caravans. They got the name Caravan as they travelled with convoys of traders. In later centuries, the dogs travelled with the army of the Nizam of Hyderabad and the soldiers used to gift the dogs to farmers who helped them when they bivouacked in the villages. It is a short-coated sighthound standing 65 centimetres tall and weighing between 22 to 28 kilograms. The coat is usually fawn coloured but it can come in varied colours and it is not uncommon to come across black coats. The jaws are long and strong. The long, sharp muzzle of the Caravan is distinctive, and one of the traditional ways of checking the purity of the Caravan breed is to see if a bangle can pass through the muzzle of the dog up to its forehead. Amol Deshpande, a breeder, says that it is a one-master dog and will obey only its handler. Anita L.G. Oechslin, an expert on indigenous dogs, points out that the Caravan and Azawakh breed of Africa are similar in nature.

A group of about forty Caravan aficionados met on 29 March 2015 at Koneri near Kolhapur in Maharashtra to take steps to protect and standardize this breed. In a letter to the international dog certifying body, the FCI (Fédération Cynologique Internationale), they pleaded for the collection of DNA samples from a wide population base to arrive at a decision on breed characteristics. The following is an extract from their appeal.

> We are deeply concerned about the survival of a most marvellous breed of dog, a living piece of India's history, the Caravan hound, also called 'Karwani', in the language of its region of origin. We address you at this critical time in the hope that you will lend your support and consider our case concerning the Karwani. From years of extended study and a recent trip to India, we are disturbed that the breed in its pure form is rapidly disappearing due to a lack of opportunity to perform its historic function and rampant, widespread cross-breeding with greyhounds, whippets,

Western Salukis, and possibly other sighthound breeds. We consider it the duty of kennel clubs to preserve and protect breeds, but feel strongly that our organizations are falling short in this regard where the Caravan hound is concerned. Through misguided and perhaps dubious breeding practices, we feel that the Caravan hound being promoted in show rings and thus being presented to the world stage differs vastly and considerably from the authentic Karwani found in rural villages in India. We find it absolutely abhorrent that while the rustic, pure Caravan hound disappears into oblivion, a modern incarnation bearing only a faint resemblance to the proud, primitive hound of old should be considered to be India's first 'breed' to be represented internationally. We present here some photographs of show-winning Caravan hounds and authentic rural Caravan hounds so that you, fellow fancier, may exercise your own judgement based on subjective observance and/or knowledge of type of any purebred dog that distinguishes one breed from another.

In the same month, another group of breeders, called the Karwani Group, met at Baramati near Pune and sent a report to the Kennel Club of India suggesting breed standards for this dog. Seasoned breeders like Sunil Pawar and Balasaheb Jachak were part of this group. There are, however, differences, among breeders about the standards to be followed, with claims and counter-claims. Nevertheless, this breed seems to be gaining in popularity at home and in the West. A company in the United States that produces T-shirts celebrating different breeds of dogs has come out with one for Caravan hounds. Beside a silhouette of the hound, the legend on the shirt says, 'I was normal until I got my first Caravan hound'.

Chippiparai

In recent years this breed had gained popularity, and the pups are sought after. W. V. Soman called it the greyhound of south India, as it is an archetypal sighthound of southern Tamil Nadu, particularly in the Ramanathapuram area. It is a quiet dog and not given to much barking. Leggy and elegant, these dogs are great chasers. Aerodynamically designed by nature, the head is long and elongated with the eyes placed right on top rather like a snipe, providing the dog with wide vision. Slim and wiry, the Chippiparai comes mostly in biscuit colour. It stands about 63 centimetres tall and weighs 30 kilograms. Siva Siddhu, popularly known as Siva Petlover on social media, has been promoting this breed. Siva points out that the Chippiparai comes in different colours and accordingly is known by different names. For instance, a specimen with a dark tan coat is called paruki. And a dog with a tan coat is called sevalai pullai. Siva says that some of its pedigree lines, such as the Seeni Nayakkar line, are well-known. Chippiparai is a small town in Kovilpatti district of Tamil Nadu, but this breed seems to have nothing to do with that place. I have not been able to find the reason for the dog's name. Their owners in the villages, particularly those who still use them for hunting, merely refer to them as pullaikanni or vettai nai (hunting dogs). This is the most popular breed of all the dogs of Tamil Nadu.

Kaikadi

Kaikadi is the name of a nomadic people who are stonemasons.

They make equipment like grinding stones for the kitchen. They also specialize in gathering herbs and making native medicines out of them. They are to be found in Maharashtra and parts of Gujarat, and the dogs they rear are called Kaikadi dogs. These dogs hunt in packs, targeting monitor lizards, mongooses and even squirrels. It is a rather small animal, standing about 40 centimetres tall and weighing around 20 kilograms. It has thin, long legs, but powerful thighs and hocks and its tail is long and tapering, with a head that is long and thin with prominent eyes and long ears that stand erect when alert. The breed has short hair that requires little maintenance. It resembles a whippet. Its coat can be white, tan or black. As a breed it is not very popular. The Kaikadi is best suited for large open areas, not urban homes.

Kanni

A sighthound, the Kanni looks distinctive, stands about 64 centimetres tall, and weighs around 35 kilograms. Built on the lines of a typical hunting dog, it has pendent ears, and a long, thin, rat-like tail. It is very rare to hear it bark. The Kanni is slim like a cheetah and is light on its feet. The coat is usually black and the muzzle and legs are brown. Occasionally, you might find an off-white dog; it is then referred to as a paala kanni (milky white kanni). Like the Himalayan mastiff, some Kanni dogs sport two distinct brown spots just above the eyebrows. Such dogs are referred to as pottu kanni, meaning 'the one with a tilak'. Dogs that do not possess these brown spots are known as karun kanni, meaning 'the black kanni'. This breed, though rare to come across, can be seen in villages along the foothills of the Western Ghats in Tamil Nadu, from Tirunelveli to Pollachi. I saw my first Kanni in

Samathur village near Pollachi. I saw another one being exhibited in the Nagercoil Indian Breeds Dog show in 2016.

The word 'kanni' means a maiden and it is said that a Kanni dog is given as part of the dowry when a girl gets married. The story may be made up but a Kanni is traditionally never sold but only gifted. Like other breeds of south Tamil Nadu, this is also used to hunt. The loyalty of this breed is legendary. There are indigenous dog breed experts who argue that the Kanni is not a separate breed but a variant of the Chippiparai. The debate continues.

Kurumalai

The Kurumalai breed occurs in the Kurumalai tract of Tirunelveli district in Tamil Nadu. Short-coated and standing about 60 centimetres tall and weighing around 30 kilograms this dog, like other sighthounds, is used for hunting hares and antelopes. A very agile dog, it usually has a blackish-brown coat. Brindle-coloured specimens can occasionally be seen. Some breeders claim that this is just another variant of the Chippiparai.

Mudhol Hound

The Deccan Plateau is home to quite a few indigenous breeds, and well known among them is the Mudhol hound. A small town in Bagalkot district of north Karnataka, Mudhol was once the seat of a tiny kingdom. One of the scions of the dynasty is credited with bringing the Mudhol breed into prominence. Raja Maloji Rao Venkatrao Ghorpade of Mudhol was fond of hunting and he always took a brace of dogs with him. During the British Raj he acquired some greyhounds and crossed them with local hunting dogs in his kennel thus creating the Mudhol. The pamphlet brought out by the Canine Research and Information Centre at Bagalkot says that one of his successors presented a pair of Mudhol puppies to King George V in 1937, and this drew attention to the breed. Subsequent kings, including Bharavasingh Maharaj of Mudhol, took a keen interest in the breed and stabilized it. In Bombay, D. J. Madan, of the well-known Madan Theatre family, took a number of steps to promote Mudhol hounds.

In the town of Mudhol, the king entrusted the care of this breed to the Chandanashiva family, and descendants of this family continue to breed them today. Shahu Maharaj, the Raja of Kolhapur, also patronized this breed and trained these dogs to assist hunters in pig-sticking. A statue of one of the former rulers of Kolhapur shows the raja astride his horse, indulging in pig-sticking with the help of a Mudhol hound. The raja, who was also fond of racing, organized dog races in Kolhapur, but this sport did not catch on. The Bedars of Gulbarga, a farming community that is much given to hunting, and the Talwars of Bijapur, also used these dogs in hunting. Its chasing prowess is said to be phenomenal, and it is reputed to be able to turn in mid-leap and change course in a split second.

The Kennel Encyclopaedia has an entry on this dog with a photograph. In the late 1980s some kennel clubs in Mumbai began to reserve a class for this breed and allow them to be shown. It was in 1990 in a show in Mumbai that a Mudhol hound was shown in the ring for the first time. Then they began appearing in show rings in Bangalore.

The Mudhol hound is built on typical sighthound lines with long legs and a curved stomach. The brisket is well-sprung and the muzzle is pointed. The head is long and the coat is usually tan, although one can also find dogs with white coats with black or tan patches. Its long tail is one striking feature of this breed. It is a large dog standing 70 centimetres tall and weighing around 30 kilograms. Bitches are slightly smaller. Temperamentally it is friendlier than many other hounds.

Dr Mahesh Doddamani, head of the Canine Research and Information Centre, says, 'Mudhol hounds are the result of cross-breeding amongst greyhounds, Saluki and Sloughi breeds of dog. When the invader arrived in our country they brought with them dogs from their lands which they used to develop the next generation of dogs by cross-breeding with local dogs.'

There is concern among many Mudhol aficionados that there are breeders who are even now cross-breeding greyhounds with Mudhols. They point out that this is spoiling the Mudhol dogs by introducing in them characteristics of greyhounds which are not suitable to the tropics, such as sensitivity to heat. True Mudhols will be unaffected by tropical conditions. Those who buy Mudhol pups need to check if they are greyhound crosses. The database prepared by the Mudhol Dog Breeders Association will be helpful in this matter.

Pashmi

A group of dog lovers in Hyderabad who fancy Pashmis are trying to save and propagate the breed. When we drove into the compound of a farmhouse in Vanaparuthi near Hyderabad two years ago, a pair of handsome dogs, looking like distant cousins of Salukis, came bounding to greet their master who was at the wheel. There was a lively spring in their gait as if they were trotting on the tips of their toes. I was told they were of the Pashmi breed and that there is revived interest in this dog.

The other name for this breed, Old Afghan hound, supports the story that it originally came from Afghanistan a few centuries ago along with the Pathans and Rohillas who came into India and settled in the Deccan. The village of Janwal in the Latur district of Maharashtra is pointed out as the place where the dogs are concentrated. Mhalangi is another place which is famous for Pashmis. Rajasaheb Patil, a farmer in Mhalangi, is one of the better-known breeders of Pashmis. He is concerned about the future of the breed. He points out that there are only about six farmers in the area who are following sound methods of breeding.

A well-fed, full-grown Pashmi stands 55 to 60 centimetres tall and weighs about 20 kilograms. They thrive in desert areas. Their bodies are well-sprung and slender, with the head held erect as is typical of a sighthound. They have arched stomachs, feathery ears, tails and legs. In fact, Pashmi means hairy, as in Pashmina shawl. (It is derived from the Persian term 'pashm' which is the fine hair under the fur of certain animals.) The hair on the ears is the hallmark of this breed. They are excellent chasers and have traditionally been used for hunting hares. A Pashmi's stamina for running is phenomenal. In flat country, it could outpace a chinkara.

The few Pashmis I have seen are grey or white. Evidently they come in other colours, including black. I have seen a rare brown one at a dog show in Bangalore. *Hutchinson's Dog Encyclopaedia* documents that these dogs were used in the Northwest frontier, in undivided India, as guard dogs. There is a first hand account of seeing these dogs on duty guarding two forts in this area. There they were referred as Baluchi hounds.

U. K. Rajwade, Deputy Inspector General of Police in the state in the 1970s, inducted Pashmis into the police canine squad in Maharashtra. I am not sure how they fared as tracking dogs. Today, the most sought after police dog is the Belgian Malinois, which established a formidable reputation in the Iraq war.

Rajapalayam

Perhaps the most valued of all the south Indian dogs is the legendary Rajapalayam. The breed is still in good shape. The average height of the dog is 65 centimetres and it weighs around 35 kilograms. It is not slim as the Chippiparai or Mudhol. Its colouring is mostly white, although occasionally one might see one that is mottled white. The eyes are brown and the pink nose is distinctive. With button ears and a whip tail, it is deep-chested and well sprung. The loose-hanging upper lips, the well-wrinkled throat and its serious expression make it look formidable. Though it has been used traditionally as a guard dog, and for hunting, it has been put to military use as well. In the Poligar Wars of the nineteenth century, some poligars used Rajapalayam dogs effectively against British cavalry. In the dark of the night, these dogs were slipped stealthily into stables where they bit through the hamstrings of the horses and put them out of action.

It would be useful at this point to briefly discuss the term poligar dogs. The earliest reference we have of poligar dogs is in Major E. Napier's book *Scenes and Sports in Foreign Lands* published in 1840. The word 'poligar' is a mutated form of the Tamil term 'palayakarar', a vassal of the Vijayanagar kingdom who was in charge of a palayam, or fiefdom. After the battle of Talikota (or Thalaikottai) in 1565 in which the combined armies of the Bahmani sultans defeated the Vijayanagar army, the palayakars had to move south, right up to the tip of Tamil country, where they established themselves as vassals of the Nayaks who ruled from Madurai. There were seventy-two palayams under the Nayaks. In times of war they supplied soldiers to fight on behalf of the raja and in return collected taxes in his area. The dogs that they had with them and nurtured were referred to as poligar dogs. My conclusion is that it is Rajapalayams that are most frequently referred to as poligar dogs. M. Krishnan writes in his book *Jungle and Backyard* about a poligar dog he raised in the 1940s named Choki. The drawing he has made of the dog and its description are certainly those of a Rajapalayam. He describes the head of Choki as having 'wrinkled dignity' and its coat as white. In an article in *The Statesman* he confirms that the poligar dog is indeed the Rajapalayam.

There is an interesting, though apocryphal, account of a poligar and his dogs. Kattabomman, the poligar of Panchalankurichi, has gone down in history as having rebelled against the British East India Company, paying for his act of rebellion with his life. The ballad telling his story includes an account of an encounter between his dogs and a hare. His brace of dogs chased a hare, and the poor animal ran for its life. After a distance, the hare turned around, stood its ground, and the dogs were perplexed. The story goes that Kattabomman, observing that the soil at this location was so empowering that it could make even a hare turn against dogs, built his fort there. The fort, of course, was razed to the ground by the forces of the East India Company. A new memorial to the rebel poligar, designed like a fort, now stands at this spot.

Though this breed is classified as a sighthound, it seems to possess considerable ability to follow a scent-trail. My friend Alagar Jagadeesan of Madurai wrote to me about his experience with this breed. When he was a boy, sixty-five years ago, his father brought home a full-grown Rajapalayam dog. The dog was friendly but when it was unleashed after an hour it disappeared and could not be found. After some time when the father visited his friend, the previous owner of the dog, in his village 80 kilometres away, he found the dog there.

Rampur Hound

The Rampur hound was one of the earliest Indian breeds to appear in the show ring. These sleek, smooth-coated sighthounds come from Rampur, a remnant of the once-proud Rohilkhand kingdom that was reduced to a small territory through successive defeats at the hands of the British during the Rohilla Wars of the eighteenth century. This tiny princedom, ruled by a nawab, came under the suzerainty of the British in 1801, retained its autonomy and a distinct cultural heritage of which the Rampur hound was a part.

The present Rampur district, situated between Moradabad on the west and Bareilly on the east, halfway between Delhi and Lucknow, was the erstwhile Rampur state. This hound has been a favourite of Indian nobility for nearly 300 years. Ahmad Ali Khan Bahadur, who was the Nawab of Rampur, is credited with developing the breed through cross-breeding with Tazis, Afghan hounds and English greyhounds. Maharaja Jaideep Singhji of Devgadh Baria, near Vadodara, also patronized and worked towards the preservation of this breed. However, according to *The Imperial Gazetteer of India*, there is a Southern connection. Volume I records (p. 183): 'Rampur is celebrated for its breed of hounds, originally introduced from south India. They are generally grey in colour, with a smooth coat, larger than greyhounds and the animals so bred are easier to train than the pure bred.'

In 1895, the periodical *The Field* carried a drawing of a Rampur hound, done by Arthur Wardle, to illustrate an article on Asiatic greyhounds. *The New Book of the Dog,* Volume II, published in 1911, carries a photograph taken in 1879 of a Rampur hound named Eileen, owned by one Lieutenant Colonel J. Garstin of Multan. This is arguably the earliest photograph of this fabled breed. However, what is to be borne in mind is that in these pictures the dog looks quite different from present-day Rampur hounds. For instance, the muzzle is not sharp and the chest is not well-sprung. Is this an indication of cross-breeding?

The dog made its first appearance in the Western world after the Prince of Wales returned from his 1876 tour of India. In the same year at the Fakenham Dog Show, two Rampur hounds featured in the show ring, and they came to be known as the greyhounds of the East. A judge at the show recorded that one was of 'a mouse colour and the other spotted, a sort of pink and blue, somewhat similar to young plum-pudding-coloured pigs'. J. Sidney Turner in *The Kennel Encyclopaedia* writes that this breed originated from Arab and Persian hounds imported into the Rampur kingdom in colonial times. He also suggests that the Rampur breed was cross-bred with English greyhounds to improve its legs.

It is a thin, large dog, standing about 70 centimetres tall and weighing around 35 kilograms. It comes in mostly brindle or mouse-grey colours, and rarely in black. The body is long, the chest very deep and the slightly arched back is distinctive. The tail is thin and long and is held low. When in action, the thin tail is carried horizontally. With its long legs and short coat it has the streamlined look of the greyhound but with a long, strong skull. The light yellow eyes give a hard expression to the dog's visage. They are set on an elongated muzzle, providing the dog with wide peripheral vision of almost 270 degrees—this makes the Rampur hound a sighthound par excellence. Owners say that this hound is slow to bark. Every year at a festival near Rampur, in which contests like chasing a hare are held, the best hound exhibited is conferred the title Rustum-e-Rampur. I learned that in the 1983 mela a rare black Rampur hound showed up.

Vaghari Hound

In 1996, I was travelling in the Saurashtra region and stayed on a farm mainly to photograph a Gir bull. The person in charge of the bull had, in addition to a striking moustache, an impressive dog, and when I asked him about it he said it was a Vaghari hound. This medium-sized dog, which stands around 55 centimetres tall and weighs around 35 kilograms, gets its name from a nomadic tribe known as Vaidava Vaghari in this part of Gujarat. The coat of my informant's dog was brindle and it was built on the classic lines of a hound. A Vaghari hound has powerful jaws compared to other Indian sighthounds. In Gujarat you can see dogs of this breed in the encampments of other nomadic people like the Rabaris.

If you have credible information and photographs about Indian breeds not listed in this book, please write to me at theodorebaskaran@gmail.com. If I am able to use your information I will do so in a future edition of the book with due acknowledgement.

EPILOGUE
THE CANINE HERITAGE OF INDIA

When I was working as Chief Postmaster-general in Gujarat in 1996, I thought I would initiate the production of a special series of postage stamps on Indian dog breeds. However, as with all initiatives involving government, I had to wait for the right people to assume office so that the idea would actually be realized. When a fellow dog lover, Barindranath Som, took over a senior position in the Postal Directorate, the time seemed ripe to push the proposal. I had retired from service by then and was in Chennai, but fortunately the Kennel Club of India, Chennai, under the leadership of C. V. Sudarsan, was keen on promoting indigenous breeds through a set of postage stamps. I worked with them, and a proposal was sent with eight photographs of the four breeds that were chosen—the Mudhol hound, Rajapalayam, Rampur hound and Himalayan sheepdog. I wanted to include the Kombai, the iconic breed of Tamil Nadu, but I could not get a good specimen to photograph. The senior officer at the Postal Board steered this proposal through the committee.

In January 2005, the Minister for Communications, Dayanidhi

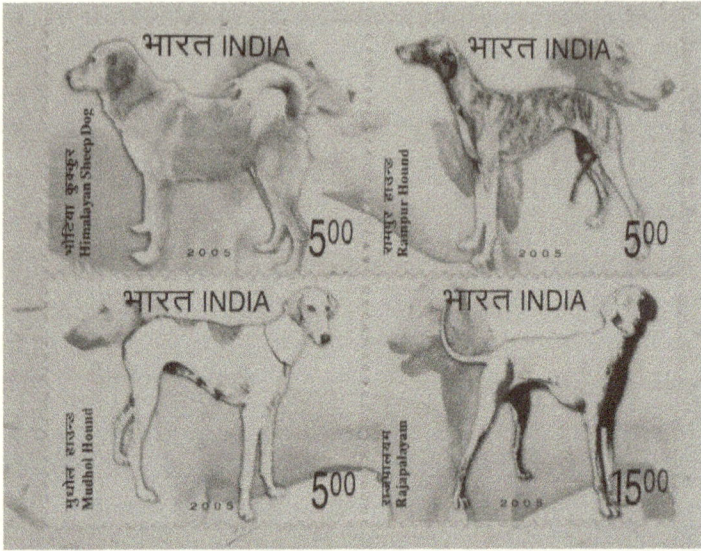

Four Indian dog breeds represented in postage stamps.

Maran, who is also a dog fancier, released four stamps at the inauguration of the annual dog show at Chennai. At the function, two thoroughbred Rajapalayam pups were given to the club to start a breeding centre. A special philately exhibition by a collector from Kerala, Tiny Francis of Tiruchur, who specializes in the theme of dogs, was installed. It was quite an attraction. I mention this initiative because as I have said throughout this book it is a shame that Indian dog breeds have been so badly neglected throughout their history. It is a matter of some encouragement that dog fanciers have banded together to spread the popularity of certain breeds but there needs to be a great deal more concerted effort in this regard.

This particular Kennel Club of India annual event in Chennai showcased some indigenous breeds which attracted considerable attention. Lynette and Peter Watson of Australia, who were on the international judging panel in the show, wrote, 'Apart from the Rajapalayam, the Rampur hound, the Mudhol hound and the Himalayan sheepdog (sometimes referred to as the Bhutan [sic]), there were also a number of other very fine Indian breeds that arrested our notice such as the Chippiparai, the Caravan hound, the Kombai and the Kanni. The Kennel Club of India is extremely focused on the promotion of the national breeds and also on their preservation.' In 1980, Doordarshan telecast a short video on Kombai dogs, shot in the villages where they are raised.

It is of critical importance that our indigenous breeds receive official international recognition from organizations like the FCI in Belgium. Although breeds like the Himalayan mastiff and Lhasa apso are popular in Western show rings, they haven't received official recognition. A proposal for the recognition of the Caravan hound has been sent to the head office of the FCI but the whole procedure may take a few years. First, the scientific commission of the FCI will have to give its approval. Once this is done a breed warden will be sent to the country of the breed to check if what has been recommended is truly a distinct breed. Details like the number of Caravan hounds registered in the last five years and the number of these hounds shown in the various dog shows in India have to be provided. The rules require pedigree details for six generations of the dog to be filed along with the proposal for recognition.

For many breeds the standards have not been clearly laid down and DNA sampling is often resorted to in order to check claims. Dr D. Krishnamurthy, Coimbatore-based show judge of international repute and vice-president of FCI Asia-Pacific, points out that breeds like the Kintamani Bali dog of Indonesia and the Taiwan dog had to wait for nearly ten years before all formalities with the FCI were completed and recognition was granted. Currently the FCI has recognized and laid down the standards for 343 breeds of dogs from around the world, including ancient breeds like the Azawakh from Saharan Africa and the Akita from Japan.

Apart from international recognition, which will help in raising the profile, breeding standards and recognition of our dogs, it is unfortunate that in our own country, government support for our canine heritage is almost non-existent. Even as breeders argue about what constitutes individual breeds, the government which could have established programmes and standards has done very little. Even fledging projects such as the one that was being carried out by the National Bureau of Animal Genetic Resources, Karnal, to survey local breeds have been shut down. In spite of filing an RTI application I could not elicit the reason for the closure of the project.

Of the state governments, only Karnataka and Tamil Nadu have shown some interest in native breeds. As we have seen, the concern shown by Karnataka for its Mudhol breed has yielded remarkable results. In February 2016, the Canine Research and Information Centre in Bagalkot presented six Mudhol pups, born in its whelping unit, to the Indian Army, which is considering inducting them into its canine wing. The army, which has a centre in Meerut, trains dogs for military duties like landmine detection and probably intends to use Mudhols for guard duty in border areas and in defence installations.

Other excellent efforts include those undertaken by the Kennel

Club of India under C. V. Sudarsan. The KCI is promoting the awareness of indigenous breeds by conducting specialty dog shows and educating rural breeders on scientific breeding. Some judges have emerged as experts in the breed characteristics of Indian dogs. There are seventy-two KCI clubs registered throughout the country and the awareness about local breeds has spread to almost all the branches. It was the KCI which spearheaded the proposal for the postage stamps featuring Indian breeds. In July 2015, the club brought out a special issue on Indian breeds in the *Indian Kennel Gazette,* the premier publication in India relating to dogs. This journal provides information about the availability of pups and sires, by running advertisements from breeding farms. Although dog shows organized by the KCI have been criticized in certain quarters for converting working dogs into show animals, there is no doubt the kennel clubs serve a useful purpose by and large and create awareness about breeds and the responsible ownership of dogs. In fact, as part of the annual dog show, KCI conducts a one-day workshop on 'Responsible ownership of dogs'.

What is heartening is the number of dedicated breeders working along methodical lines to protect and increase the popularity of certain breeds. Upender Reddy of Hyderabad is an advocate for the Pashmi, Prasad Mayekar of Mumbai specializes in the Caravan hound, and Siva Siddhu in Tamil Nadu is promoting the Chippiparai. Mayekar has nearly fifty Caravan hounds in his kennel and is a regular figure at specialty shows. Some breeders have set up farms and are managing them on commercial lines. From all accounts these breeds are gaining popularity.

Social media organizations like Facebook provide channels through which breeders and dog fanciers keep in touch with each other and make an effort to raise awareness on indigenous breeds. As the message slowly spreads, we see new converts to Indian breeds in the dog shows. It is also encouraging that in many dog shows prizes have been instituted in a new category—Best Indian Breed.

But a lot more work needs to be done. First, it is crucial that indigenous breeds are recognized as being an important part of India's heritage. When this is done it will make it easier to set up breeding programmes for select breeds, and extol their merits to dog fanciers. This might also serve to reverse the mania for the import of unsustainable foreign breeds into the country. It is not too late to protect our indigenous breeds but we need government and institutional support to provide the foundation on which individual efforts can be based. Our unique breeds deserve nothing less.

APPENDIX
THE MATTER OF STRAY DOGS

Over the last fifteen years we have seen periodic reports in the press about stray dogs and their increasing population. This is a recent phenomenon. Until two decades ago, the population of stray dogs, pigs and monkeys in India were kept under control through legal measures. In Tamil Nadu, for example, municipalities were authorized to destroy ownerless animals like dogs and pigs. One of my childhood memories is of visiting the municipal office in my hometown to get licence tags for our dogs as unlicensed animals were rounded up by municipal dog units and euthanized. But the dog licence regulation, like many other civic rules that made our lives safer in the past, has fallen into disuse.

The Animal Rights Movement began to spread in India in the 1980s and one of its aims was a law to prevent the euthanizing of stray dogs. The Prevention of Cruelty to Animals Act 1960 was invoked and the elimination of stray dogs and pigs was stopped. The Government of India (oddly enough, through the Ministry of Culture which has nothing to do with animals) issued the Animal Birth Control (Dogs) Rules, 2001 (ABC for short), through which a scheme was launched to spay and castrate stray dogs with a view to check their population. However, very few of the provisions of this set of rules are being implemented. In fact, it is difficult to implement them. The provision that people have complained about is that the dogs are to be released in the same place from where they were caught, after they have been castrated. For instance, the plan talks about forming local committees to implement the rules, but this has not been done. The scheme, which has been in operation for over two decades, has not made any dent in the stray dog population or in the incidence of rabies. The World Health Organization (WHO), in a bulletin on rabies, said the incidence of the disease in India has been constant for a decade. It goes on to say that this is probably an understatement, as rabies is not a notifiable disease in India. (If it were a notifiable disease, all hospitals, including private ones, would have to mandatorily report to the government any case of rabies that they come across.)

Consider the facts: rabies or hydrophobia, the deadliest of all infectious diseases, and a major public health hazard, claims 35,000 lives in India each year and sadly half of them are small children. That is 81 per cent of all the rabies deaths in the world, according to a WHO report. This lethal virus is passed on to humans mostly by stray dogs. (There are no stray dogs in Lakshadweep and no rabies either. Sikkim has also eradicated rabies.) According to one estimate, every two seconds someone in the country is bitten by a dog. Many of these dogs are rabid dogs. Most of the victims are poor, and 40 per cent are children. The annual dog bite incidence in India affects an astounding 17.4 per cent of the population. In 2007, the cost of treatment of dog bite cases in India was a whopping ₹750 million. Some years ago, I witnessed the death of a friend's five-year-old son from rabies in Tiruchirapalli. He was bitten on the head, and even immediate anti-rabies shots could not save him. My interest in the subject began at that time. I have had anti-rabies shots twice—fourteen pokes around the navel in those years—and for twenty days went to work wearing a loosely tied dhoti.

The rabies virus attacks the central nervous system and kills the victim in a few days through convulsive seizures. Once infected, death is certain. The problem with rabid dogs is that in the initial days of the illness, they appear perfectly normal and stay close to humans as usual. Often, by the time one realizes that a dog is rabid, the damage is already done.

So how does one deal with the problem of stray dogs? Remember India is home to thirty million stray dogs, the highest population in the world, not just in absolute numbers, but the highest in proportion to the human population. In our cities there is a stray dog for every thirty-five humans. To give you a few examples of just how widespread the problem is, let's review the situation in some of our cities. In Patna, around 2,000 dog bite cases are reported every month by government hospitals. Dr B. R. Benjamin, scientist at the Indian Veterinary Research Institute, Hyderabad, recorded in *People's Reporter* that a survey done in Hyderabad, and Secunderabad, revealed that at any given time, there are 12,000 rabid dogs spreading pain and death. Approximately 350 to 400 dog bite victims report daily at the Preventive Medicine Institute, popularly known as Rabies Clinic. A survey done in Delhi revealed that in a period of six months, in 2014, between June to December, 8,783 dog bites were recorded. According to the last official survey of stray dogs conducted by the Municipal Corporation of Delhi in

2009, the population of stray dogs was approximately 5.62 lakhs.

It has been calculated that there is one death by rabies every thirty minutes in India. According to the calculation of an NGO, in the city of Bangalore there are two lakh stray dogs. India is one of eleven nations where the risk of contracting rabies is the highest in the world. You see posters to this effect in travel agencies in Europe.

In 2013, Mission Rabies started functioning in fourteen countries with the aim of eradicating rabies from the world by 2030. India, which is considered the world's hotspot for rabies, is one of its major concerns and it has permanent vaccination teams based in Goa and Ranchi. Professor M. K. Sudarshan of the Rabies Epidemiology Unit, Kempe Gowda Institute of Medical Sciences, Bangalore, speaking at the Second Commonwealth Veterinary Conference in Bangalore in February 2000, pointed out that most of the rabies victims in India are the poorest of the poor who have no access to anti-rabies vaccine. These include workers who return home late, people who live on dimly-lit streets and children who walk to school. The cost of treatment of each dog bite victim is ₹1,500. India's estimated need for anti-rabies vaccine for humans is 1,500 litres per year. Since all the vaccine produced in the twelve centres in India fall short of our requirement, we resort to importing the vaccine from France and Germany. Sudarshan pleaded that as a first step the public must be educated about stray dogs and this dreadful disease.

◆

Is it possible to contain the population of stray dogs by neutering them? The Indian Veterinary Research Institute in Hyderabad, had done a census of stray dogs in the twin cities of Hyderabad and Secunderabad, and the figures show that spaying is a pointless exercise. It has to be borne in mind that dogs are quite fecund and have two breeding seasons in a year. It has been calculated that one breeding pair would have given birth to 382 pups in the third year and 12,680 pups in the fifth year. By neutering a few dogs in cities, we do not even scratch the surface of the problem. One may feel satisfied for having done something good, but that is all. Let's take a look at the cost: it costs ₹600 to remove the uterus from a bitch; that does not include the cost of post-operative care. Moreover, who takes care of the dog after surgery? A veterinarian friend tells me that 50 per cent of the strays that are operated on die due to lack of post-operative care. Further, an anti-rabies vaccine needs to be given once a year to each dog, if it is to remain protected. Can this be done to the millions of strays?

Simply put, attempting to neuter all the country's strays is like trying to empty a vast lake with a bottomless bucket. Unless we operate on all the dogs, male and female, in one go—an impossible proposition—the scheme is bound to fail. As a Tamil proverb goes, you cannot jump across half the well. Making matters worse is the fact that it is not just the strays that make up the population of ownerless dogs. In an act of utterly irresponsible dog-ownership, some pet-owners abandon old pets and pups in public places like parks. In Bangalore, during littering season, it is a common sight to see dozens of pups abandoned in Lalbagh Garden and Cubbon Park. The ones that survive join packs. Therefore, what is the point in neutering a few?

And what about rabies? The WHO recommendations for the control of dog population and rabies incidence are given in a single sentence: 'Promote and enforce pet control laws, undertake sustained re-immunization and eliminate unwanted dogs.' India's national policy for dog population and rabies control called the ABC programme, developed by the Animal Welfare Board of India (AWBI), has been implemented in various cities since 1994. The programme is centred on the neutering of unclaimed homeless dogs and putting them back on the streets. This programme concentrates on and directs all infrastructural and financial resources on that population of dogs that do not belong to anyone, cannot be caught annually for re-vaccination and whose welfare is compromised by the very fact that they are homeless.

The first step in tackling the stray dog problem is to accept that animal birth control is totally ineffectual. Some other effective way to reduce the population of stray dogs must therefore be worked out. Many countries have begun to figure out that there is probably only one truly effective way. In June 1997, following cases of rabies in Taiwan, the government organized the largest mass extermination of stray dogs in history. Seven lakhs of them were destroyed in a matter of days. Countries that have traditionally set the standards for dog care, like the United Kingdom, destroy strays. If a dog is not claimed in two days, an inspector of the Royal Society for the Prevention of Cruelty to Animals puts the dog to sleep.

In 2002, an NGO called Stray Dog Free Bangalore filed a complaint with the Lokayukta who said people who were responsible for eliminating the menace were taking a casual approach to the problem. Various city corporations in the country have been trying legal remedies to tackle the problem. In 2009, the Bangalore City Corporation went to the Supreme Court on the issue. The Special Leave Petition filed told the Supreme Court that the Animal Birth Control scheme on which ₹14 crore had already been spent was 'a wholly unsuccessful scheme'. The petition stated that the ABC was neither scientific nor holistic, as it covered just a tiny fraction of the stray dog population and sought to strike it down. The petition stated that 'the rules seek to protect animals, with complete disregard to what is demanded for the safety of human beings'. The Supreme Court in 2015 announced that it would revisit the laws relating to stray dogs and then decide.

Rabies is not the only threat from stray dogs. In 2015, in Hyderabad, a four-year-old girl named Swapna, the daughter of a labourer, was ripped apart by a feral pack, even as people watched. There are periodic reports of instances of newborn babies being snatched away by strays from crowded government hospitals. Strays are also the source of a disease that is transmitted to humans from the faeces of dogs. Add to this, the number of road accidents caused by stray dogs. Writing about this problem Janaki Lenin and Romulus Whitaker said, 'Even snake-bite is of minor importance in comparison.'

Another issue that has compounded the problem concerns the phenomenal decline in the vulture population all over the country. In the last fifteen years the use of the drug diclofenac in the treatment of cattle has been found as the cause for this decline. Vultures, which fed on the carcasses of the dead cattle, died. A survey found that in the absence of vultures, stray dogs move in and feed on these carcasses and their numbers have increased.

In recent years, another problem from these ownerless dogs has attracted the attention of scientists—the threat to wildlife. Two wildlife biologists, Suman Jumani and Arjun Srivastha of the

National Centre for Biological Sciences, Bangalore, drew attention to this issue after observing five dogs chase and kill a spotted deer at Maravakandy Dam near Mudumalai Sanctuary in Tamil Nadu. Such incidents have been reported from many places across the country. In the Indira Gandhi National Park in Chennai, stray dogs frequently prey on the fawns of chital and blackbuck. Such encounters between stray dogs and wildlife can have disastrous consequences. In villages near Sholapur in Maharashtra there were wolf attacks on people, and examination revealed that these animals were rabid. They had contracted rabies from stray dogs in that area. Jumani and Srivastha point out, 'mission phenomenon may have repercussions beyond our current understanding'. Wildlife expert Pritam Chattobadhyaya has written in *Conservation India* about stray dogs preying on blackbuck at Vetnai. He concludes: 'The romanticism of Jack London's *Call of the Wild* evaporates fast when one considers the feral dog scourge in the country. I have witnessed several occasions where feral dogs are still being pampered despite creating serious social and ecological problems. Man's best friend is not wildlife's.' On the eastern coast of India, particularly in certain areas of Odisha, Andhra Pradesh and Tamil Nadu, Olive Ridley sea turtles—declared endangered—come ashore in winter to lay eggs. Dogs roaming these beaches dig out the nests and feed on the eggs. Observations by the Andaman and Nicobar Environmental Team show that 90 per cent of the highly endangered Leatherback sea turtle nests are destroyed by stray dogs. This widespread predation has pushed this largest of the sea turtles to the very brink of extinction. H. S. Singh, member of the National Board of Wildlife, has in a recent paper titled 'Stray Dogs: A Major Threat to Wildlife' said that stray dogs are a serious threat to blackbuck in Gujarat. He points out that in Mehesana district the blackbuck population has fallen from 5,000 to 1,000 mainly due to predation by stray dogs. Singh adds in this article that the dogs also kill the common fox and the desert fox in the Wild Ass Sanctuary in the Little Rann of Kutch.

Stray dogs attacking wild asses in the Rann of Kutch.

◆

In mid-2015, the problem peaked in Kerala. The number of dog bites reported in the year 2013-2014 was 93,000 and it rose to 184,000 in 2014-2015. People across the state began protesting. Certain local bodies (municipalities, panchayats) resorted to putting down strays, and some functionaries justified the killing.

In the town of Cannanore dogs were found dead, presumably poisoned by unknown people. What happened in the town of Thrissur was symptomatic of the mood of the public. Kochouseph Chittilapilly, a businessman, went on a hunger strike in a public place demanding action to end the stray dog problem in the town. He announced a donation of ₹5 lakh to any group that euthanized ownerless dogs. He started a social media campaign by placing a YouTube clip showing the last painful moments of rabies patients. He is continuing his campaign on many fronts. Meanwhile, a public interest legal case is being tried in the High Court of Kerala. In another case, the Supreme Court of India expressed concern over the rising 'stray dog menace in Kerala'. It expressed concern that children and pedestrians were bitten by stray dogs. Appearing on behalf of Janaseva Sisubhavan, a Kerala-based organization working for the cause of street children, lawyer V. K. Biju said that children become easy targets for dogs. He submitted that 'Kerala is changing from "God's Own Country" to "Dog's Own Country".' The apex court asked the state government if it could collect all the stray dogs in one place and sterilize them.

This has grown into a public health and environmental concern of immense magnitude and the situation is growing worse with each breeding season. Clearly, none of this is the fault of the dogs themselves, and one wishes that methods such as ABC were effective. But sadly, without decisive action, this public health and safety issue will not soon disappear.

Introduction

1 **'Before humans milked cows, herded goats or raised hogs...
they had dogs':** James Gorman, 'The Big Search to Find Out
Where Dogs Come From', *New York Times*, 18 January 2016.

2 **research project carried out in Sweden:** Pontus Skoglund,
Erik Ersmark, Eleftheria Palkopoulou, Love Dalen, 'Ancient Wolf
Genome Reveals an Early Divergence of Domestic Dog Ancestors
and Admixture into High Latitude Breeds', *Current Biology 25*, 2015.

3 **'Dogs may have been domesticated much earlier':** 'Dogs
May Have Split From Wolves 10,000 Years Earlier Than Thought',
Archaeology, 22 May 2015.

3 **In 2011...the carcasses of two puppies:** Anna Liesowska, 'Ancient
puppy's brain is "well preserved"... as dog bares its teeth after 12,400
years', *Siberian Times*, 16 March 2016.

3 **...a team of scientists presented their findings on the
origin of the dog:** Laura M. Shannon et al., 'Genetic structure
in village dogs reveals a Central Asian domestication origin', *PNAS*,
19 October 2015, 10.1073/pnas.1516215112.

5 **Database of the DNA of ancient dogs:** Gorman, 'The Big
Search'.

Section I: The History of Dogs in India

10 **A study done in Adelaide, at the Australian Centre for Ancient DNA, has linked ancient Indian canine visitors:** 'The Big Search to Find Out Where Dogs Come From', *The Conversation*, 15 January 2013.

12 **a hunting scene depicting a dog walking along with a man:** T. S. Subramanian, 'Discovering and deciphering rock art', *Frontline*, 27 November 2015.

16 **The East India Company officials requested four dogs:** Isaac Job Thomas, *Paintings in Tamil Nadu: A History*, Oxygen Books, 2014, p. 240.

Section II: The Contemporary Scene

32 **silver vessels for these dogs:** J. L. Kipling, *Beast and Man in India: A Popular Sketch of Indian Animals in their Relations with the People*, London: Macmillan and Co., Limited, 1904.

33 **greyhound racing at Faridabad in Punjab:** Ellen Barry, 'Chasing the Lure of a Royal Past with Greyhound Racing in Punjab', *New York Times*, 9 January 2016.

33 **banned the import of foreign breeds:** T. M. Cinthya Anand, 'Indian summer not for exotic dogs', *The Hindu*, 29 April 2016.

37 **Bangalore has seen about fifty dog spas:** Abhineet Kumar, 'Ratan Tata now funds love for dogs', *Business Standard*, 5 January 2016.

38 **there are 1,432,522 pet dogs in this city:** *Times of India*, 16 December, 2015.

Section III: Guide to Indian Dog Breeds

53-54 **provides a graphic description of how Indian hunting dogs were used:** M. Krishnan, 'World of Smells', *The Statesman*, 21 April 1979.

63 **a rare golden-coated Tibetan mastiff was sold by a breeder**

for $2 million: Parismita Goswami, 'Tibetan Mastiff Puppy Sold for $2 Million in China', *International Business Times*, 20 March 2014.

78 **concerned about the survival of a most marvellous breed of dog, a living piece of India's history:** 'The Karwani'. Report from the Karwani Group to KCI, March 2015.

83 **dog with a photograph:** J. Sidney Turner, *The Kennel Encyclopaedia*, 1908.

87 **poligar is indeed the Rajapalayam:** M. Krishnan, 'The Indian Country Dog', *The Statesman*, 19 June 1983.

Epilogue

93 **there were also a number of other very fine Indian breeds that arrested our notice:** *Madras Musings*, 1 March 2005.

94 **wait for nearly ten years before all formalities with the FCI:** Interview with the author, 6 September 2015.

Appendix: A Matter of Stray Dogs

98 **81 per cent of all the rabies deaths in the world:** Dr K. Sandeep, 'Man's Worst Foe', *The Hindu*, 6 May 2002.

98 **every two seconds someone in the country is bitten by a dog:** Read it here: http://www.missionrabies.com/.

99 **how does one deal with the problem of stray dogs?:** *Current Conservation*, 8 October 2014.

99 **India is home to thirty million stray dogs:** Neetu Chandra, 'Rabies stalks India with its 30 million stray dogs', *India Today*, 6 April 2014.

99 **there are 12,000 rabid dogs spreading pain and death:.** *People's Reporter*, 16 July 1998.

99-100 **According to the last official survey of stray dogs:** 'No census on street dogs in last six years', *Indian Express,* 17 August 2015.

100 **there is one death by rabies every thirty minutes in India:**

Times of India, 25 February, 2016.

100 **the public must be educated about stray dogs:** M. K. Sudarshan, 'Assessing the Burden of Rabies in India', at the WHO Rabies Survey, 2004.

101 **one breeding pair would have multiplied into 382 in the third year**: *The Humane Society of the United States.*

102 **Stray Dog Free Bangalore files a complaint:** *The Hindu*, 30 June 2002.

103 **in the absence of vultures, stray dogs move in:** Nikita Mishra, 'Alert! Vultures on the brink of extinction in India', *The Quint*, 30 June 2015.

103 **road accidents caused by stray dogs:** Janaki Lenin, 'Dogs and Us', *Indian Express*, 14 January 2007.

104 **five dogs chase and kill a spotted deer**: Sumana Jumani and Arjun Srivathsa, 'When domestic dogs are used for hunting', *The Hindu*, 6 December 2012.

105 **stray dogs are a serious threat to blackbuck**: Himanshu Kaushik, 'Did dogs wipe out 80% of Kadi's blackbucks?', *Times of India*, 25 October 2015.

106 **Kerala is changing from 'God's Own Country' to 'Dog's Own Country':** Krishnadas Rajagopal, 'SC expresses concern over rising stray dog menace in Kerala', *The Hindu*, 1 March 2016.

BIBLIOGRAPHY

'Dogs Domesticated Over 27,000 Years Ago: Study', *The Hindu*, 23 May 2015.

'Nature in Literature, Art, Myth and Ritual', Volume 8, No. 1, *Pandanus,* 2014.

'Study links ancient Indian visitors to Australia's first dingoes', *The Conversation*, 15 January 2013.

'The Karwani'. Report from the Karwani Group to KCI, March 2015.

Baker, Samuel White, *The Rifle and the Hound in Ceylon*, London: Longman, Brown, Green and Longmans, 1854.

Anand, Cinthya, 'Is it goodbye to foreign breeds in Bengaluru homes?', *The Hindu*, 29 April 2016.

Barry, Ellen, 'Chasing the Lure of a Royal Past with Greyhound Racing in Punjab', *New York Times*, 9 January 2016.

Baskaran, Theodore, S., 'Canine Watch', *The Hindu*, 9 January 2005.

Belsare, Anirudha, 'Diseases of free-ranging dogs: Implications for wildlife in Conservation in India', *Current Conservation*, 8 November 2014.

Bhakat, M., 'An Ode to the Bakharwal Dog', *Dogs & Pups*, Jan-Feb Issue 2010.

Bradshaw, John, *In Defence of Dogs,* London: Penguin, 2011.

Bukhari, Shujaat, 'Fear of Bakerwali dog going extinct', *The Hindu*, 16 November 2011.

Burton, R. W., Lt. Colonel, 'Days and Doings with my Bobbery Pack', in *Natural History and the Indian Army*, J. C. Daniel and Lt. Gen. Baljit Singh, eds., Bombay: Bombay Natural History Society, 1886.

Corbett, Jim, *The Man-Eating Leopard of Rudraprayag*, London: Oxford University Press, 1948.

Dayma R. G., & B. A. Gadgil, 'Police Dogs', *Science Age*, November 1986.

Debroy, Bibek, *Sarama and Her Children*, New Delhi: Penguin, 2008.

Desai, Rishikesh Bahadur, 'Tombs for "Saintly" Parrots and "Royal" Dogs in Bidar', *The Hindu*, 29 November 2015

Dhyanesh, Jathar, 'Dogged Determination', *The Week*, 20 June 2015.

Forsyth, J, Captain, *The Highlands of Central India: Notes on their Forests and Wild Tribes, Natural History and Sports,* London: Chapman and Hall, 1889.

Francis. F. ICS., *The Madras District Gazetteers*, Madurai, 1906.

Fryer, John, *A New Account of East-India and Persia*, London: Rose and Crown, 1898.

Ghosh, Amitav, *Flood of Fire*, New York: Farrar, Straus, and Giroux, 2015.

Gorman, James, '15,000 Years Ago, Probably in Asia, the Dog was Born', *New York Times*, 19 October 2015.

Gorman, James, 'The Big Search to Find Out Where Dogs Come From', *New York Times*, 18 January 2016.

Harari, Yuval Noah, *Sapiens: Brief History of Humankind*. London: Penguin, 2011.

Hubbard, Clifford, *The Afghan Handbook*, London: Nicholsons and Watson, 1957.

Hultzsch, E., ed., *Epigraphia India*, Volume VI, New Delhi: Archaelogical Survey of India, 1901.

Hutchinson, Walter, *Hutchinson's Popular & Illustrated Dog Encyclopedia*, London: Hutchinson & Co, 1935.

Indian Kennel Gazette, Indian Breeds Special issue, July 2015.

Kannaiyan, V., *Scripts In and Around India*, Madras: Government Museum Bulletin, 1960.

Kipling, J. L., *Beast and Man in India: A Popular Sketch of Indian Animals in their Relations with the People*, London: Macmillan and Co. Limited, 1904.

Krishnamurthy, S., *Hero Stones*, 2004.

Krishnan, M., 'World of Smells', *The Statesman*, 21 April 1979.

Krishnan, M., *Jungle and Backyard*, New Delhi: National Book Trust, 1961.

Krishnan, M., 'The Indian Country Dog', *The Statesman*, 19 June 1983.

Kumar, Abhineet, 'Ratan Tata now funds Love for Dogs', *Business Standard*, 5 January 2016.

Lenin, Janaki, 'How Dogs have played a part in killing wildlife', *FirstPost*, 31 July 2011.

Lockwood, M. C., 'A Mystery Dog in Sculpture', *Indian Express*, 6 March 1976.

Lockwood, Michael, ed., *Indological Essays*, Chennai: Madras Christian College, 1992.

Lorenz, Konrad, *Man Meets Dog*, London: Methuen & Co., 1954.

Messerschmidt, Don, *Big Dogs of Tibet and the Himalayas*, Bangkok: Orchid Press, 2010.

Mosley, Leonard, *The Last Days of the British Raj*, London: Weidenfeld & Nicolson, 1961.

Nagaswamy, R., ed., *Seminar on Hero Stones*, Government of Tamil Nadu, State Department of Archaeology, 1974.

Napier, E. Major, *Scenes and Sports in Foreign Lands*, London: Henry Colburn, 1840.

Nelson, J. H., *The Madura Country: A Manual*, Madras: Asylum Press, 1868.

Osborn, Lt. Gen., 'Indian Sheep Dogs', *Natural History and the Indian Army*, J. C. Daniel and Lt. Gen. Baljit Singh, eds., Bombay: Bombay Natural History Society, 2009.

Pandian, Thomas, *Indian Village Folk: Their Work and Ways*, London: Elliot Stock, 1897.

Reddy, Upender, 'Recognising and Loving Our Indian Treasure Forever', *Dogs & Pups*, 20 November 2009.

Right to Information Act, 2005. Information collected through No. NBAGR/RTI Act/application/2014–15/714 dated 29 September 2015.

Sarpotdar, Mrinalini, 'The Dog in Ancient and Medieval India', *Science*

Age, Vol. 4, No.8, November 1986.

Sayeed, Vikhar Ahmed, 'The Hounds of Mudhol', *Frontline*, 26 June 2015.

Schimmel, Annemarie, *The Empire of the Mughals: History, Art and Culture*, Chicago: University of Chicago Press, 2006.

Shahu, Chhatrapati (Maharaja of Kolhapur), Vilas Adinath Sangave, B. D. Khane, 'Rajarshi Shahu Chatrapathi papers Rajarshi Shahu Chhatrapati Papers: 1910-1913', Kolhapur: Shahu Research Centre, 1994.

Shannon, Laura M., et al. 'Genetic Structure in Village Dogs Reveals a Central Asian Domestication Origin', *Proceedings of the National Academy of Sciences*, 3 November 2015.

Siddu, Siva, *Complete Study of Chippiparai/Kanni Breed* (circulated monograph), 2015.

Sidney Turner, J., ed., *The Kennel Encyclopaedia*, London: George Routledge & Sons, 1908.

Sinclair. W. F., 'Notes on Indian Breeds of Dogs', *Journal of the Bombay Natural History Society*, No. IV, 1892.

Skoglund, Pontus; Ersmark, Erik; Palkopoulou, Elethefheria and Dalén, Love, 'Ancient Wolf Genome Reveals an Early Divergence of Domestic Dog Ancestors and Admixture into High-latitude Breeds', *Current Biology*, Volume 25, Issue 11, p 1515–1519, 1 June 2015.

Soman, W.V., *The Indian Dog*, Bombay: Popular Prakashan, 1963.

The Imperial Gazetteer of India, Volume I, Oxford: The Clarendon Press, 1909.

Thomas, Mini P., 'Unsound Bite', *The Week*, 12 July 2015.

Tiwari, Deepak, 'Man's Best Debt', *The Week*, 4 January 2015.

Vacek, J., 'The Dog in Sangam Literature', *Pandanus '06. Nature in Literature and Ritual*, 2006.

Von Holdt, Bridgitt M. et al., 'Genome-wide SNP and Haplotype Analyses Reveal a Rich History Underlying Dog Domestication', *Nature*, 8 April 2010.

Wade, Nicholas, 'New Finding Puts Origins of Dogs in Middle East', *New York Times*, 17 March 2010.

Warmington, E. H., *The Commerce between the Roman Empire and India,*

New Delhi: Munshiram Manoharlal Publishers, 1928.

Weber, Andreas, 'Best Friend', *Geo*, October 2012.

Welsh, Colonel James, *Military Reminiscences: Extracted from a Journal of Nearly Forty Years' Active Service in the East Indies*, London: Smith, Elder and Co., 1830.

Yong, Ed, 'A Genetic Study Writes a New Origin Story for Dogs', *The Atlantic*, 19 October 2015.

PHOTO CREDITS

I am grateful to the following artists, photographers and libraries for allowing me to use their wonderful pictures in *The Book of Indian Dogs*.

Boopathy Srinivasan: Colour sketches of the Alaknoori, Banjara, Kaikadi, Kurumalai and Sindhi.
Sukhjinder Singh: Colour photos of the Caravan hound, white Pashmi, Mudhol, Rajapalayam.
Karthik Davey: Colour photo of three Chippiparais.
Nithila Baskaran: Colour photo of Himalayan sheepdog with Changpas in Ladakh.

Job Thomas: Pages 13, 15.
Kalyan Varma: Page 105.
The Estate of M. Krishnan: Page 39.
M. Muthukrishnan: Page 11.
Pundalik Dhuri: Pages 24.
Roja Muthiah Research Library: Page 4.
The Department of Archaeology, Tamil Nadu: Page 18.

INDEX

Bakharwal

Himalayan mastiff

Himalayan sheepdog with Changpas in Ladakh.

Jonangi

Kombai

Koochee

Sindhi

Pandikona

Patti

Lhasa apso puppies for sale at Bomdila, Arunachal Pradesh.

Tibetan spaniel

Tibetan terrier

Alaknoori

Banjara

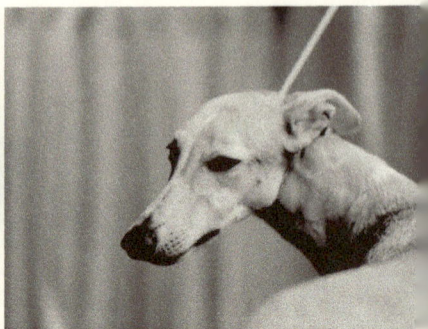

Caravan hounds

Chippiparais

Kaikadi

Kanni

Kurumalai

Mudhol

Pashmis

Rajapalayams

Rampur hounds

Vaghari hound